jets

jets

Airliners of The Golden Age

James Ott and Aram Gesar

Airlife
England

ISBN 1 85310 183 4

First published in 1993
by Airlife Publishing Ltd.

Text written by James Ott
Edited by Michael O. Lavitt and Christopher Fotos

All photographs:
Copyright © 1977-93 Aram Gesar
Except on pages:
British Aerospace: 7, 11, 13-16, 84
British Airways: 19, 26-30
Japan Airlines: 23
McDonnell Douglas: 101
Swissair: 22

Jet Transports Characteristics Section
Compiled and Edited by Aram Gesar
Three-Drawings Copyright Pilot Press Ltd.

A Book Conceived, Edited and Produced for Airlife Publishing by
Pyramid Media Group
116 Hudson Street
New York, NW 10013

Copyright © 1993 Pyramid Media Group, New York

All rights reserved. No part of this publication may be reproduced or
used in any form or by any means – graphic, electronic, or mechanical,
including photocopying, recording, taping or information storage and
retrieval systems – without written permission, except in the case of
brief quotations embodied in critical articles or reviews.

Printed by Kyodo Printing Co. (S'pore) Pte Ltd.

Airlife Publishing Ltd
Shrewsbury, England

Introduction

Jet-powered commercial aircraft and the people who work with them have changed our world. In slightly over three decades since their introduction, jet airliners have become the primary means of international travel. When jet-powered transports superseded propeller-driven aircraft, speed was their forte and the reduction in travel time their great appeal. For a transatlantic sojourner in a jet, Europe or North America became a six- to seven-hour non-stop trip, rather than one of 12 or 13 hours. The slower prop planes soon fell to the jet on the grounds of economics, convenience and novelty.

Jets have made this flying age possible, or at least hastened its arrival. Their reliability has contributed to an improving level of safe operation, even as the industry grows. The record of the General Electric CF6-80 series of high-bypass engines provides a prime example of reliability. Consider these facts:
- Some 2,894 engines were in service in 1987 on 851 airplanes of all wide-body types.
- Early versions of the engine had a high time record of 42,296 hours in service.
- The engines flew an average of 8.1 hours a day and were involved in 2,840 departures a day.
- They were ready for flight 99.9% of the time.
- In-flight shutdown rate was .02 per 1,000 engine flight hours. The CF6-80C2 version had no in-flight shutdowns in 1986.

In recent years, an airline ticket has come within the financial reach of many people. Improving technologies for airframes and engines have added the drive for economy. The trend is toward two less-noisy, more fuel-efficient engines instead of three or four. Computer-aided automation has made possible the design of a two person cockpit. Governments have taken liberalized attitudes toward aviation, freeing it from restrictive regulations and recognizing its place as foremost in the mass transportation system. Marketing techniques and competition among airlines have given further impetus to low fares.

Critics of these trends have said service levels have fallen as jets have become popular. This certainly has been true in the United States, where airline deregulation was introduced in 1978, and it has now spread to the Far East, Canada, Europe, Australia and Chile. Cabin service on short-haul flights has deteriorated in the United States, or is non-existent today. The new Business Class featuring roomy seating and fine foods has been a welcome addition and a relief from Economy Class. Service in the First Class cabin has improved by a good measure, recalling the elegance of the early days of the Jet Age, as airlines have expanded into the international market.

Military operations have first claim on the jet as a powerplant. Jet fighters had been in world arsenals for more than a decade before the first commercial jet transport arrived in the 1950s with the British Comet. Our presentation ends more than three decades later with the Airbus Industrie A340, the newest and most technologically advanced commercial aircraft. We include of course the Anglo-French Concorde, the only supersonic airliner

Concordes typically require 7,500 feet (2,225 meters) of runway to land.

that, though the technologies are from the 1960s, still provides for us a glimpse of the aircraft of tomorrow.

The epoch between the 1950s and the 1980s will be known as a golden age of commercial aircraft production. It was a time when each airline's engineering department had as much or more to do with designing an airplane as the manufacturers. Airlines were profitable, they knew what they wanted, and the manufacturers filled those needs. Today there are few developmental engineers with the airlines. Aircraft design is undergoing a revolution of its own in the decade of the 1990s. The Boeing Co. in coming years will introduce the first "paperless" airplane, the Boeing 777. The aircraft is being created by computer in three dimensions in a pre-manufacturing stage. This preview should ease the integration of the many systems and components that go into an aircraft and eliminate the many conflicts that can muddle the manufacturing stage.

Boeing has assembled design-build teams from its engineering design office, its manufacturing sector, from finance and support, and crucially, from the 777's customers and the support sector. These teams are reviewing the impact of the design prior to finally committing the aircraft to manufacture.

But the jets of a golden era are our focus. We hope to bring them alive by the recollections and observations of airmen and airwomen.

The Jet Engine

The jet engine wedded to a commercial airframe found a harvest in the era immediately after World War II. Advances of the 1930s and the 1940s were put to use:
- More powerful jet engines for speed.
- Swept-back and thinner wings for supersonic flight.
- Lift-improvement devices for short field operations.
- Weight savings.

Speed and convenience – the appeal of the modern age – made air transportation attractive. In the United States, railroads competed by hiring French chefs to prepare dining car cuisine. New and more luxurious passenger cars were placed in service, but the appeal and advantages of jet-powered airplanes eclipsed the railroads as the primary mode of mass transportation between U.S. cities in the 1950s.

The growth of airline passenger traffic begun in this post-war period continued until 1991, when a combination of concerns about terrorism and the ill-effects of the Persian Gulf War caused traffic to decline below the previous year's levels for the first time in history. The passenger traffic decline was exclusive to 1991, although the financial impact has extended for several years.

Airlines were profitable in the post-World War II era, and they began looking for aircraft that would meet the needs of a growing network of routes. Manufacturers, flush with new ideas from war work, were able to satisfy the needs of the airlines. It was the beginning of a golden age of transport aircraft manufacture. The jet engine, conceived but barely developed before the war, saw limited service in the conflict and was developed to its potential after the war. The debt to the military for advances in engine and airframe technology is obvious.

"Traffic growth was exploding. Airlines were making money. It was a golden age for air transport."
– Joseph Sutter, Boeing Commercial Airplanes, retired, on the Jet Age.

Coincident with the Boeing 707 was the USAF KC-135. Before the wide-body transports there was the Lockheed C5 and the high bypass jet engine. Advances in engine efficiency and reliability came almost hand in hand with less noisy, less polluting engines.

With the jet transport, the pilots, the designers, the maintenance crews and the empire-builders created a transportation system. Air service in the United States began taking the form of non-stop city pairs. By 1970 in the U.S., more than 1,000 city pairs were being connected by nonstop flights.

A 747 starts a take-off run.

Air passenger miles (passengers times the miles flown) exceeded total miles traveled in Pullman railroad cars in the United States for the first time in 1951. The air miles were 10.67 billion, compared with 10.22 billion ground miles.

In less than two decades, beginning in 1950, airlines of the world upgraded their powerplants from reciprocating (propeller-driven) engines to ever-more advanced new jet powerplants. The jet era for air transport was inaugurated by the turbojet-powered de Havilland Comet 1A, which flew its first flight on July 27, 1949. The turbofan engine, with its greater efficiencies, power and fuel economy, launched the wide-body jet era with the Boeing 747. Its first flight was on February 9, 1969.

The fuel efficiency of the turboprop jet engine and the potential of turbofans for high subsonic speeds has led to research into propfan engines. While propfans are fuel efficient, airlines have been hesitant to move to this new technology for larger passenger aircraft. Noise and vibrations were early problems with the propfans. These engines are easily recognizable for their propellers, which look like the blades of a kitchen blender.

In the jet engine lay the answer to the quest for greater speed. Jet engines overcome the aerodynamic problems associated with propeller-driven aircraft operating at speeds approaching Mach 1. Propellers become less efficient. Jet engine intakes are designed to accept the air and flow it through the engine.

Other aircraft structures, particularly the wing, require changes for efficiencies in the high subsonic or transonic speeds. One characteristic of early jet aircraft was the swept-back wing, which enables the aircraft to approach Mach 1 speeds without the buffeting and control problems of a straight wing. In later models, devices to increase lift capabilities were applied to aid short-field capabilities

Early Development

The names of a handful of people are associated with early jet engine development. The American Sanford Moss, head of the turbine research department at General Electric, developed a turbo-supercharger that was applied to the reciprocating engine on a LaPere biplane in 1920. This device compressed air going to the intake of the engine and enabled it to operate at high altitudes. The LaPere aircraft broke records, gaining an altitude of more than 33,000 ft. A French patent for a turbojet engine was obtained by Guillaume in 1921.

On January 16, 1930, a 23-year-old Royal Air Force officer, later Sir Frank Whittle, filed British Patent No. 347,206. This filing was a description of a practical turbojet. Sir Frank signed "FWhittle" with a flourish on the drawing of his engine that was to revolutionize military and commercial aviation. He received private support for his venture in 1935, and his engine was successfully bench tested in April, 1937.

Whittle's engine was not the first jet to power a flying aircraft, however. This accomplishment belongs to Hans Joachim Pabst von Ohain, who joined the German aircraft manufacturer Ernst Heinkel in 1936. Bench test of the von Ohain engine was completed in September, 1937. Heinkel manufactured an aircraft, the He 178, as a test bed for the von Ohain engine. The He 178 inaugurated jet-powered flight on August 27, 1939.

Austrian-born Anselm Franz built the Jumo 004 engine for Junkers Engine Co. in Germany. It powered the Messerschmitt Me 262 fighter in World War II.

Reciprocating engines had reached their pinnacle. Propellers lose efficiency as they approach Mach 1. Other aircraft components – the engine cowlings, the fuselage, the canopy, the trailing edge control surfaces – are affected as well at transonic speeds, the result of air compression associated with the higher speeds. The compressibility effects on the aircraft of the time are legendary – buffeting, lack of stability and control. Manufacturers began to design aircraft around the new engines to avoid or minimize these effects. Earlier aircraft had been designed on the theory that air was not compressible. The new jet aircraft slowly took on streamlined looks.

Below: Four de Havilland Ghost turbojets awaiting a Comet 1.

Below – Top: Front view of the Comet.

Bottom: Back of the Rolls-Royce Conway turbojet that powered early Boeing 707-420s.

An Airbus A300, followed by a Boeing 747-200, taxis along Runway 13R/31L at New York's John F. Kennedy International Airport.

Jet Aircraft

BRITISH AND FRENCH INNOVATORS

The de Havilland Comet

Great Britain emerged from World War II victorious and poor, but its aviation industry was intact. As a nation it had the edge in jet engine technology, and among its aeronautics people, there was keenness to exploit the advantage for commercial purposes. Thus, the sleek and innovative de Havilland Comet, later produced by Hawker Siddeley, was developed. Its maiden flight took place on July 27, 1949.

Four de Havilland Ghost turbojets, stashed in the wing roots, powered the Comet transport. Each engine developed 5,000 lb. of thrust. By today's standards, the Comet 1A was a poor performer. It was capable of 490 mph, or Mach 0.74 at 35,000 ft. It could carry a maximum payload of 44 passengers in a pressurized cabin to a range of 1,750 miles. But in its early days, the Comet was exceedingly popular with passengers. Sensing the coming triumph of jet power, airline managers took lively interest.

The Comet's wings were swept back at a modest 20 degrees, but otherwise the aircraft looked much like the advanced propeller-driven transports of the 1950s. Placement of the engines, two in each wing root, sparked debate. On one hand, the engine alignment close to the fuselage contributed to the longitudinal center of gravity. If one engine were to fail, the yawing motion would be held to a minimum. The position in the wing, however, complicated maintenance, and some critics thought it unwise to locate engines there for fear of passenger harm and critical damage to the aircraft or controls from an uncontained engine failure.

". . . It wasn't until the jet engine came into being and that engine was coupled with special airplane designs – such as the swept wing – that airplanes finally achieved a high enough work capability, efficiency and comfort level to allow air transportation to really take off."
– Joseph F. Sutter, Executive Vice President, Boeing Commercial Airplanes, in the 23rd Wings Club General Harris 'Sight' Lecture

Below: The British Comet rolls out, launching a new era in commercial aviation.

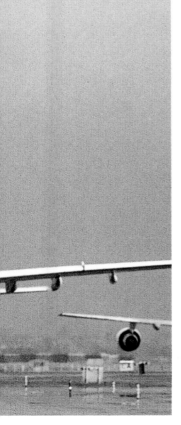

Another issue was raised by pilots over the Comet's take-off thrust-to-weight ratio, which was a low 0.17. The aircraft required precise handling during takeoff to avoid over-rotation, substantial drag effects and the loss of acceleration.

The Comet inaugurated pure jet passenger service on May 2, 1952, on the London-Johannesburg route. Then tragedy struck when a Comet broke up in flight on January 10, 1954, and a second Comet disintegrated in the air on April 8, 1954. The Comet's unanticipated problem lay in an improper design that did not predict the stresses from cabin pressurization cycles. The fleet of 20 Comets was grounded in 1954 as engineers dissected the aircraft. The problem was dubbed metal fatigue.

But the Comet continued to triumph through the 1950s, introducing jet service to numerous markets around the world. Modified and re-engined versions were developed that initiated transatlantic service. Rolls-Royce Avon turbojets powered later models. British industry built 74 re-engined Comet 4s. But the tragic loss of aircraft ruined the Comet for wide and continued acceptance.

A derivative model of the Comet is the Nimrod, approved for the Royal Air Force in June, 1965, for surveillance, antisubmarine warfare, and early-warning duties.

Britain's Canadian associates broke new ground by developing the first commercial jet-powered airliner in North America. The aircraft was the Avro Canada C-102 Jetliner, which flew on August 10, 1949, two weeks after the Comet. Two Rolls-Royce Derwent 5 engines were paired on each wing. Test flying of the 50-passenger airliner attracted no customers, and the project was abandoned.

In the Soviet Union, the Tupolev Tu-104 was developed along the lines of military aircraft. The 50-passenger jet entered service with Aeroflot in 1956.

Above: Comet 2X in flight with Rolls-Royce Avon 502 engines, showing the enlarged air intakes. This improved version of Comet first flew on February 16, 1952.

Below: Test equipment inside the Comet 1, the first jet airliner.

The Comet's African flight schedule in the early days of jet travel.

One of the Comet 2s built for the Royal Air Force.

Left and below: Comet engines found a home inside the wings.

Left: The Comet 4C became the final production version.

The Caravelle

The mounting of its two engines, one on each side of the aft section of the fuselage, marked the French Sud-Est Aviation S.E. 210 Caravelle as a revolutionary commercial transport.

This positioning of engines established a precedent that was to influence design of twin and tri-jets of the future.

A Caravelle prototype flew for the first time on May 27, 1955, powered by Rolls-Royce Avon turbojet engines.

Air France placed orders for 24 Caravelles in December of that year. The short-range jet entered Air France service on July 27, 1959, on a route to the United Kingdom. Air France continued Caravelle service until some 22 years later, March 28, 1981.

> **"The Caravelles were well-built structurally. They had excellent handling qualities, but they were not that easy to work on. There were a lot of access panels to get at what you wanted to and the fasteners were not quick-release."**
> – Clyde Kizer, former United Airlines test pilot and head of maintenance for grounded Midway Airlines, currently president of Airbus Service Co., Inc., Herndon. Va.

Airlines operated the Caravelles in seating configurations ranging from 60 to 80 passengers. The range at maximum payload was more than 1,800 miles, with twin-engine economy.

United Airlines operated a fleet of 20 Caravelles. Ralph Glasson, retired vice president for maintenance at United, recalled an unusual feature of the aircraft, at least unusual to the United officials who were evaluating it.

"The initial design was such that it had lights in the cockpit that told you things were going correctly," Glasson said. "We were used to red lights when things were not right. But this was an initial concern and it was corrected before deliveries to United."

There were no red "all-right" lights in United cockpits.

In this period inter-airline cooperation had reached a high point. Scandinavian Airlines System and Air France both had experience operating the Caravelles, and both carriers provided technical assistance to United.

Another characteristic of the Caravelles was the nose design, which followed a pattern close to that of the de Havilland Comet. Caravelle designers worked on the Airbus A300, the twin-engine wide body that launched the European consortium Airbus Industrie, in the aircraft manufacturing business. In the next decade, learning from the Caravelle experience and their own brilliant grasp of design, these skilled hands went on to great achievement in the A300 as a passenger and cargo aircraft, according to United's Glasson.

Production of the Caravelles ended in the early 1970s after 280 airplanes of various versions were built. In 1988, some 62 Caravelles were still in passenger and cargo service. Anti-noise legislation was limiting their usefulness to those nations that were not signatories to international anti-noise standards.

A Caravelle 10R powered by Pratt & Whitney JT8D turbofans, one of the last versions produced. Rolls-Royce Avon turbojets powered the second version of the Caravelle III. The Caravelle cockpit borrowed its design from the Comet. The noses are also similar.

THE AMERICAN RESPONSE
Boeing Bounds Ahead

As the British Comet made newspaper headlines for its speed and luxury, The Boeing Company in the early 1950s was designing the aircraft that would project it to pre-eminence in the commercial field. The experimental ship was called the 367-80, the Dash 80, the prototype for the Boeing 707 commercial jet. The aircraft is known in United States military parlance as:
- KC-135, the series of aerial tankers;
- E-3, Airborne Warning and Control System (AWACS);
- E-4 aerial command pos;
- E-6 Tacamo, an airborne strategic communications relay platform;
- Air Force One, the president's airplane, until the late 1980s.

In civilian or military guise, the Boeing product is one of the company's outstanding financial successes and an enduring technical achievement.

Boeing's A. M. (Tex) Johnston piloted the yellow-and-brown prototype on the first flight on July 15, 1954. Johnston is known in the industry for executing, at another time, a slow roll of a Boeing 707 before a stunned audience of company and U.S. aviation officials at Seattle. On the second prototype test flight with Johnston at the controls, the Boeing 707 had an impromptu race with a North American F-86 Sabrejet.

> **"An F-86 near it (the Boeing 707 prototype) had a tough time keeping up with the big transport. Some performance!"**
> – **An account of the first Boeing 707 prototype trials in *Aviation Week* July 26, 1954**

The Boeing 707 owes its development in part to the military contracts for the KC-135 – a research and development debt that Boeing's foreign competitors frequently mention. Its powerplant had military origins. The four Pratt & Whitney JT3s on each Boeing 707 were civilian derivatives of the military J57 engine. Thrust of the initial turbojets was rated at 10,000 lb. The addition of a two-stage fan increased the thrust. Later JT3 models were rated at 19,000 lb. Engine mounting on the Boeing 707 is similar to that on the B-47 bomber: the JT3s hang from pylons that jut forward, ahead and below the 35-degree swept wing.

A pattern for fuselage width (140 inches) established in the Boeing 707 has carried through in the 727, 737 and 757 transports. Each of these single-aisle transports offers six-across seating, three on each side of the aisle. The characteristic, bullet-shaped Boeing nose, common to all company transports except the 747 and 757, came first with the 707.

From both design and technical standpoints, the 707 established precedents. The Boeing 707-320B's maximum gross weight is 336,000 lb., three times the weight of the British Comet. It carries as many as 189 passengers. Maximum range is 5,750 miles. The thrust-to-weight ratio of early 707 models was such that they required a 10,000-ft runway for take-off at high gross weight. Through the use of such devices as the trailing-edge slotted flaps and leading-edge flaps, which increased lift, the wing stall speed was rated at an acceptable 121 mph. Use of spoilers – the plate-like devices that rise on the wing at command and diminish wing lift – provided the Boeing 707 with a high rate of descent.

Airline service was inaugurated by Pan American World Airways in the fall of 1958. The Boeing 720, a limited-range domestic version of the Boeing 707, competed with the domestic Convair 880s and 990s. Boeing received 999 orders for the Boeing 707/720 transports, and 981 were delivered. By the late 1980s, approximately 240 707s remained in service, not counting an additional 14 Boeing 720s. A lasting tribute to the aircraft design is the selection by the U.S. Navy of the Boeing 707 as the strategic communications Tacamo aircraft 30 years after the prototype's first flight.

Opposite: A 707-320B. The 707 ushered the United States into the jet transport era and set a standard for fuselage width that was to be imitated in the 737, 727 and 757.

Below: This 707-420 was powered by Rolls-Royce Conway 505 turbojets.

Close up of the nose of a Boeing 720B. The 720 was the short to medium-range version of the Boeing 707, a response to the Convair competition.

The Convair 880 and 990 transports brought a new element of class and technology to the field. The General Electric CJ-805-3 engines were civilian versions of the GE J-79 engine. The aircraft featured Mach bumps, also known as Whitcomb bumps, which reduced the intensity and drag of shock waves.

Mr. Hughes' Transports

Convair 880s and 990s, trim and classy representations of the optimistic 1950s, were designed as transports for the elite Jet Set of the time. The programs lost $430 million as reported by the company in 1963, the year in which deliveries of original orders were completed. Sixty-five Convair 880s were produced, and 37 of the upgraded, follow-on 990 Coronados were built. A handful of each type remains 25 years later in storage, with perhaps a few operating worldwide.

The Convairs were the pet project of Howard Hughes, the wealthy aircraft designer and owner of Trans World Airlines. The 880 was so named because of its capability of covering 880 feet in a second.

"Thirty years ago I was in command of a Thai Airways International CV 990 Coronado that flew from Hong Kong to Bangkok, from takeoff to landing, in a new record time of one hour and fifty-four minutes, breaking the previous record of two hours and two minutes. Not only does this record still stand but is indeed approximately twenty minutes faster than the normal time taken to fly the route today."
– Capt. Hans Fugl-Svendsen, at the Aviation Week Pacific Rim Conference, October 16-18, 1990 in Hong Kong.

In their day, the 880 and the 990 were among the hottest commercial aircraft in the world fleet. Cruising speed of the 880 was 615 mph. A Trans World Airlines Convair 880 reached a maximum ground speed of 849 mph between Columbus, Ohio, and Pittsburgh, Pa, on Jan. 24, 1961. The Convair 990's cruising speed was 640 mph. In tests, pilots pushed it to the limit, Mach 0.98, closing on the speed of sound. (The speed of sound is 761 mph. at sea level. At an altitude of 35,000 feet, the speed of sound is about 660 mph.) Swissair pilots likened the experience at the controls of the 990 Coronado to Formula 1 race car driving.

The 990's 40-degree swept wing and design for speed gave it the characteristic of tending to roll at low speeds. Swissair Captains Bruno Schmitt and Guido Schaefer remember the aircraft as being structurally rigid. The aircraft was so stiff that landing it became a controlled crash, according to Captain Schaefer.

In the cabin the Convair transports sought a comfort level that is now unknown in commercial aviation. The aircraft's first-class interior was standard. An optional setting offered a 12-place seating area grouped around coffee tables.

Powerplants for the 880 were the General Electric CJ-805-3 engines, versions of the military J-79. The sea level static thrust rating of the 3B version was 11,650 lb. Half the engine thrust could be reversed for braking purposes.

Delta Air Lines acquired the Convair 880 and operated the aircraft in the early 1960s.

In the beginning of Delta's scheduled service, which began May 15, 1960, the cabin was entirely first class.

> After Delta's founder, Collett Everman (C. E.) Woolman, agreed to buy the Convair 880, he and Howard Hughes exchanged telephone numbers. Hughes gave two numbers, and Woolman noted that his number also was listed in the Atlanta telephone book. He asked Hughes only not to call his home after 9 p.m. On one occasion, Hughes called at 9:04 p.m., and spent the first 25 minutes apologizing.

The 990s were powered by CJ-805-23B fanjets, which incorporated an additional turbine and fan in a single stage behind the basic engine for greater efficiency. The modified engine was housed in a pod almost six feet in diameter. Typical thrust was 16,050 lb.

The wing of the 990 also was advanced for its time. Its swept-back characteristics and anti-shock bodies were designed to permit a continuous air flow over the wing's upper surface, which reduced the intensity and drag of shock waves. These Mach bumps or Whitcomb bumps – two on each wing – are easily identifiable characteristics of the Convair 990.

Swissair Captain Schaefer attributed the 990's tendency to roll at slow speeds to the sharply swept wings. Observers said the Convair aircraft were faster than their contemporaries, slightly smaller and burned too much fuel getting to speed. It was uneconomical against the McDonnell Douglas DC-8. The Boeing 720, which was Boeing's direct challenge to the Convair 990, weighed 11,000 lb less.

The original 990 was certified by the Federal Aviation Administration in December, 1961. Type certification of the international 990A came in October, 1962 and of the domestic 990A in January, 1963, the year the program was abandoned.

A Convair 990A of Swissair in flight. The airline named the aircraft "Coronado." Only 37 of this model were built.

Douglas DC-8

The cockpit of the DC-8 is a typical first generation jet transport layout with seats for the captain, first officer, engineer and sometimes for a navigator.

The Douglas Aircraft Co. of Long Beach, California, immensely successful with its string of propeller-driven transports, ran third in the competition to produce a jet-powered transport. The four-engined, swept-wing DC-8 was nearly a replica of the early Boeing and Convair jet transports. Until the middle 1960s brought about the stretch version of the DC-8, which elongated the fuselage and wings like taffy and added 60 seats, few could tell the aircraft apart.

Douglas operated as if running third were virtuous. Its idea, perhaps since it was already late, was to take more time to build a newer aircraft incorporating the latest concepts in design. The first five DC-8 versions in the series have the same dimensions. The long-range intercontinental versions – Series 30, 40 and 50 – were still the same length but had larger fuel tanks, new engines and a beefed-up structure.

First announced on June 13, 1955, the DC-8 took to the air for the first time on May 30, 1958, about six months past the original schedule. Sales were inhibited before the middle of 1958 because of the lack of a flying prototype. Production models of the Boeing 707 were flying regular airline schedules by the time the DC-8 Series 10 was certified by the Federal Aviation Administration on August 31, 1959. Other DC-8s, the Series 10 through 50, were certified by May, 1961. Between 1957 and 1961, 47 DC-8s were sold, compared with 170 Boeing 707s.

Delta Air Lines inaugurated service with the DC-8 on September 19, 1959, a scant few hours before United Airlines put its first DC-8 in the air. Delta took advantage of its flight schedule in the earlier time zone (EST) to be first. The inauguration of DC-8 service by Delta is regarded by many at that carrier as an early example of a success that gave it a lead over its competition – not with United in this case but old Eastern Airlines, Delta's historic competitor. Eastern did not buy the DC-8.

The DC-8 program spent $7 million between June, 1955 and April, 1958 on wind-tunnel tests for an improved subsonic wing. Douglas varied the camber of the leading edge, used double-slotted flaps for improved lift and produced a 30-degree, swept-back wing.

The structure of the DC-8 was built not to one safety standard, as was required, but to two, with the aim of making the aircraft:
• Fail-safe. This means that if any one member of a structure fails, the weights and pressures, known as loads, would shift by design to another member.
• Fatigue-resistant. Basically, this is a standard of overbuilding strength into the materials.

Use of the lightweight and strong metal titanium saved an estimated 945 lb. in the aircraft. Titanium was used in engine pods and pylons, and in places where special strength was needed – the tabs that are riveted into the fuselage known as rip stoppers; and the extra plates over the doors, known as door doublers.

Douglas delivered 556 DC-8s. The stretched Series 61 and 63 offer seating capacities for more than 250 passengers and international ranges. Re-engined to meet tough new noise criteria and for better fuel efficiency, many of these aircraft – named Series 71 and 73 – are still operating. In 1988, 100 of the DC-8-70 series were in service, the majority as freighters.

Above: The nose of a DC-8-71 is shown. The aircraft was designated a 71 series after re-engining a 61 series with General Electric/Snecma CFM56 engines. The re-engined DC-8 yielded a 23% improvement in fuel efficiency, compared to the JT3D-powered aircraft.

Left: A McDonnell Douglas DC-8-61 lands. Its Pratt & Whitney JT3D turbofan engines are the DC-8's original powerplant type.

The McDonnell Douglas DC-8 comes in many versions and re-engined models are still flying the skies. The aircraft is remarkable in that it was built to two safety standards: fail-safe and fatigue-resistant. The DC-8 was one of the first commercial aircraft to use titanium for strength and weight savings.

The Vickers VC10 is powered by four Rolls-Royce Conway engines.

MORE FROM BRITAIN

The Vickers VC.10

Britain kept trying to exploit its edge in jet engine technology. The Vickers VC.10 began to take shape at Weybridge in 1959 after a crucial management decision. As a second-generation transport, the VC.10 could have focused on speed. Instead, the design emphasis lay on airfield performance and economy. Yet it was no slouch in the speed category as a 600-mph. transport.

Taking the lead from the Caravelle, Vickers engineers mounted the four engines, two by two, at the rear of the fuselage. The horizontal stabilizer shifted upward for a "T"-shaped tail. Maximum gross takeoff weight of the long-range VC.10 approached 300,000 lb. It was designed to carry the full payload of 38,000 lb. nonstop between Mexico City and New York.

The VC.10 is remembered as a British transport par excellence. Peter Adams, currently a first officer with Swissair, flew the right seat of the VC.10 for what is now British Airways. He recalls the VC.10 as offering elbow room for the cockpit crew. The cockpit had space for five seats, one each for the pilot, first officer, flight engineer, navigator and for a check pilot or observer.

Pilots loved flying the VC.10 and some speak of air races with the Convair 880/990s. It was a durable British product designed for the Old Empire routes and short fields in remote areas of the globe.

A total of 54 VC.10s were built for commercial service, and none operates today.

This Super VC10 version had a lengthened fuselage and a seating capacity for 187 passengers. VC10s were faster than 707s, and Fowler flaps enabled pilots to make gentle approaches.

The Trident

A short-range companion to the VC.10, the Hawker Siddeley Trident achieved success. First known as the de Havilland 121, the aircraft was stretched, and its 3B version added long range to its characteristics. At the end of the 1980s, some 26 Tridents continued to fly around the world, including the People's Republic of China. A total of 117 were produced.

The Trident looks much like the Boeing 727. It is a trijet with two engines mounted on either side of the fuselage and the third engine in the fuselage below the "T" tail. Among pilots the Trident is known for its pioneering of actual blind landings. The test Trident aircraft, G-ARPB, made hundreds of landings in heavy fog at London's Heathrow Airport in the latter part of 1966. The Blind Landing Experimental Unit at the Royal Aircraft Establishment at Bedford performed the automatic landings, using a Ferranti inertial platform in the aircraft and a Standard Telephone & Cables localizer antenna.

Left: The Trident 3B fin contrasts with the open access panels of the Rolls-Royce RB162 boost engine access.

Opposite: The Hawker Siddeley Trident is shown in flight, water vapors trailing.

Below: The Trident had a three-man crew. A fourth seat was provided for an observer.

BEA (British European Airways) acquired 24 Trident 1Cs, the first version of the aircraft.

THE HIGH AND THE WIDE

Airlines in the 1960s perceived a great vista of passenger traffic growth. They had vanquished the railroads as the primary mode of commercial travel in the United States. They were making money. A boom in aircraft buying lay ahead, since acquisition of new transports usually follows a cycle of airline profits.

World airlines hit a stretch of remarkably good times in the middle 1960s, setting the stage for the buying spree. Bolstered by the optimism of a growing industry and coupled with an extraordinary "can do" attitude from the manufacturing sector, airline developmental engineering departments began sketching out the airplanes of the future.

Basic design of the 747 was influenced by Pan American World Airways and its head, the legendary Juan Trippe.

For a five-year period starting in 1964, world airlines achieved record-breaking earnings. In 1964, earnings reached $612 million, and in subsequent years the airlines reported earnings of $900 million, $1.037 billion, $934 million and $550 million. The run of record profits fueled interest in new, improved aircraft to accommodate the forecast high levels of traffic growth. Source: International Civil Aviation Organization statistic

Passenger and cargo traffic in this period rose at a tremendous rate. Traffic for scheduled airlines in the Western world more than doubled between 1962 and 1967, from 129.7 billion to 273.3 billion revenue passenger kilometers. And in spite of oil crises and economic downturns, by 1980, world passenger traffic had grown by a factor of 10.

A traffic jam in the airways and at airports threatened even in the 1960s. Air traffic controllers took to "stacking" airplanes when they were too busy, or when approaches to airports were at capacity. This meant that airplanes were orbiting airports at different altitudes awaiting clearances to land. (Stacking is still used today, but only when other means of managing air traffic flow are exhausted. Air traffic controllers prefer to hold aircraft on the ground, not in the air.)

Concerned with the possible abrogation of safety levels, the U.S. Federal Aviation Administration set up special rules for aircraft operations at busy airports. The FAA effectively put a lid on airplane operations at certain airports, and allocated landing slots, which airline committees dispersed among the air carriers. This situation still exists today at four major U.S. airports: Washington National, New York's La Guardia and John F. Kennedy and Chicago's O'Hare.

For the airlines, bigger airplanes were the answer. With the government restricting the number of flights, bigger transports could handle more people. Together, airlines and the manufacturers set out to build airplanes tailored to the needs of the markets that were economical to operate as well.

Opposite: The airlines wanted an aircraft with three times the capacity of the Boeing 707. Boeing launched a $2 billion program to produce the 747.

Below: The 747 changed the way that people travel by offering a wide-bodied comfort level still unmatched in the commercial skies.

Designs ran in several directions:
• The Boeing Co., urged on by Juan Trippe, Pan American World Airways' indomitable chairman, conceived of the Boeing 747 wide-body Jumbo Jet.
• Franklin W. Kolk, head of American Airlines' design bureau, foresaw a wide-body jet powered by two engines. Kolk's idea resulted 10 years later in the Airbus Industrie A300 wide-body twin.

The more powerful, fuel-efficient high bypass turbofan jet engine coming into the market in the mid-1960s made possible the flight of heavy and long-range wide-body jets. The high bypass turbofans were developed for greater fuel efficiency and thrust. The low bypass turbofans had been improvements over the ordinary turbojets that powered the first commercial jets. The differences between the jet engines deserve an elaboration.

In a turbojet the air is sucked into the compressor. The squeezed (or compressed) air is forced through the engine to a combustor where fuel is ignited. The forced air and the exploding fuel blast through a turbine and blow out the exhaust. The process produces thrust very similar to the exhaust of air from a balloon. The air flows quickly from the balloon, pushing it forward at the same time.

Engine makers in the late 1950s produced a low bypass jet engine by adding a fan to the front of a turbojet. The fan forced more air into the compressor, but most of the air went around and outside of the compressor, bypassing it. Twice as much air went around the compressor as through the engine core. The fan gave the engine more power and greater fuel efficiency.

In high bypass engines developed for the wide-body transports, the ratio of bypassing air was more like 4:1. Today's engines go up to a 6:1 bypass ratio.

The fuel efficiency advantages were obvious. Turbojet aircraft exhibited a specific fuel consumption (akin to miles per gallon for automobiles) of approximately 1.0 nautical miles per pound of fuel burned. The first turbofans produced a 15% improvement in fuel efficiency and second-generation, high bypass engines improved on that by 20-25%. Contemporary engines are 40% better.

The 747 was launched in a team effort involving Pratt & Whitney, but now all major engine manufacturers have produced powerplants for the aircraft.

First-class passengers are seated on the main deck below the cockpit in most airline configurations. Passengers view the world through large portholes in elegant spacious surroundings.

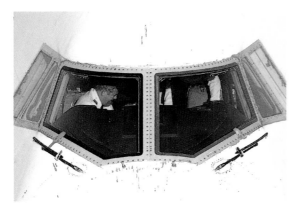

Some models have been operated for more than two decades. On-going research into the stresses of operating the juggernaut has resulted in structural work in Section 41, around the 747's nose.

Right: The rear bulk cargo door open for loading. The door's latching system was redesigned after it caused the fatal crash of a Turkish Airlines DC-10-10 near Paris

Below: The DC-10 windshield and side panel windows provide excellent visibility to the crew.

A FRATERNAL TRIJET TWIN
The Lockheed L-1011

Opposite: The L-1011 became the last commercial production aircraft of the Lockheed Corp. Early models will have been in service for more than two decades.

The Lockheed Corp., maker of the Air Force C-5A, entered the wide-body competition with the Lockheed L-1011 trijet. Its first flight took place on November 16, 1970, and it entered service in 1972 behind the DC-10.

Like the DC-10, the Lockheed TriStar arranged two of its three engines under the wings, and placed the third engine in the rear. The L-1011 engine is mounted in a fashion similar to the fuselage engine on the Boeing 727. A large inlet hole can be seen just above the aft fuselage line. An S-shaped duct links with the engine, which exhausts at the tail cone. The DC-10 engine is integrated with the tail fin, held perpendicular to it, like a large cigar.

The L-1011 is powered by Rolls-Royce RB.211-524 turbofan engines, which each are rated at 48,000 lb. thrust.

Below: An L-1011-1 set for landing. The trijets offered wide-body comfort for the long-range passenger. Their economy and efficiency have continued their value into the 1990s.

In performance, the L-1011 is similar to the DC-10. One distinction is the TriStar's short field capability, at least in comparison with the DC-10 and the Boeing 747. At maximum load, the L-1011 can operate on an 8,070-ft. runway. The DC-10 requires a 10,370-ft. runway, and the 747 a 10,450-ft. runway.

A Series 500 long-range version of the L-1011 was introduced in the early 1980s. It is an aircraft ahead of its time, the first to use fly-by-wire technology – all electric – for the activation of ailerons. But only 50 of the Series 500s were built. A total of 249 TriStars were produced before the program ended in 1983, ending Lockheed's commercial transport making business.

The ahead-of-its-time avionics suite, the fly-by-wire design and the handling characteristics were cited in an interview with several British Airways pilots as reasons for their great affection for the L-1011, especially the -500 Series. They did not, however, appreciate the layout for engine instruments and dials.

Opposite: The L-1011-500 is the shorter-bodied, long-range version with more powerful engines.

Opposite: L-1011s are powered only by Rolls-Royce RB211 turbofans. The rear engine mounting is similar to the mounting on a Boeing 727. The inlet hole can be seen just above the aft fuselage. AN S-duct links with the engine, which exhausts at the tail cone.

Above: In 1972 a Lockheed crew flew the first fully automatic coast-to-coast flight in the U.S. on an L-1011 "hands off" from takeoff through touchdown, demonstrating the aircraft's automatic flight control and autoland systems.

SUPERSONIC TRANSPORTS
Anglo-French Concorde

The supersonic Concorde represents the great technical achievement in commercial aviation of recent decades. Its success is a tribute to its British and French engineers and designers. Its lack of commercial success is the handiwork of the first fuel crisis and heavy resistance from environmentalists and politicians in the United States.

Some say the opposition was fortuitous, or the world would have had a U.S. entry, a second fuel-guzzling SST during the first world fuel crisis in 1973, and no airline with any money to buy either one.

It is remarkable to many that only 20 Concordes were built, 16 of them in production and only 14 for sale. Secondly, only two airlines fly them – British Airways and Air France. Oh, for the days when the world airlines held options for 74 Concorde transports, including a string of United States airlines, European and Far Eastern flag carriers. The outlook for sales ran into the several hundreds. These halcyon days were before the U.S. Congress killed funding for the American supersonic transport and anti-noise advocates blocked roadways in New York to disrupt the Concorde's arrival.

Opposite: Five thousand hours of testing went into Concorde, making it the most tested aircraft in aviation history. British Airways and Air France each operate seven Concordes.

Below: Concorde's tail contains a fuel tank. Presence of fuel in the tail tank in supersonic flight helps to keep the nose high.

"Nevertheless, with all of its anticipated and unanticipated operational and economic problems Concorde is now coming on strong as a contender for the Blue Ribbon traffic of the mid-1970s and already is the strongest challenger to American domination of the international transport market in history."
– Aviation Week & Space Technology editorial by Robert Hotz, February 8, 1971.

The operation of the Concorde is a standing rebuke to those who killed the U.S. program. Research work in Britain and France started on the program in 1956. Discussions between the British and French governments led to a 1962 agreement to design, develop and manufacture the aircraft jointly. Five years later a prototype rolled out at Toulouse, the corporate home of France's Aerospatiale.

Today, British Airways and Air France each operate seven Concordes. One pre-production aircraft is at British Aerospace Filton, the second at Toulouse. Both nations manufactured a single prototype, one pre-production and seven production aircraft. Development costs totalled more than $1.5 billion.

A Concorde speed record was achieved on December 30, 1985, when tail winds helped it achieve 1,490 mph. between New York and London, beating the previous record of 1,470 mph. Concorde's cruising speed is Mach 2, 1,350 mph.

Concorde was slow in coming to commercial aviation. Its first supersonic flight took place on October 1, 1970, and Mach 2 was exceeded for the first time on November 4, 1970. British Airways ordered five Concordes in 1972. In 1975, the Civil Aviation Authority in Britain awarded Concorde a certificate of airworthiness. Commercial service started between London and Bahrain on January 21, 1976.

The aircraft structure is made of aluminum alloy, along with titanium and steel used in the powerplant and landing gear components. The design service life is 45,000 hours – 25,000 hours of these at elevated temperatures associated with high-speed flight. Special features of the Concorde include:

• Four Rolls-Royce/Snecma Olympus 593 engines, each rated at 38,000 lb. of thrust, equipped with afterburners that add fuel to the final stage of the engine. The afterburners produce extra takeoff power and the extra push to supersonic flight.

• Concorde's cruise altitude is between 50,000 and 60,000 feet, 10,000 to 30,000 feet above normal cruising altitudes of other commercial jet airliners.

• Concorde's takeoff speed is between 200 and 215 knots, compared with a 165-knot takeoff speed for subsonic airliners.

• An air intake system is made specially for the Olympus engines. It uses a variable ramp assembly and a diffuser to compress the air and retard the flow into the engine to subsonic speed.

Concorde's effect on the environment was tested in Britain over an 800-nautical-mile flight corridor along the west coast of Scotland, over the Irish Sea and along the coasts of Wales and Cornwall. Little adverse effect was shown on the daily life of the estimated 1.5 million people beneath the corridor.

Opposite, top: Concorde's droop nose with its visor down. The visor is raised for supersonic flight to protect the windshield from temperatures that reach 248F (120C).

Opposite, bottom: Cockpit of Concorde with the visor up.

Right: The long, slender fuselage and smooth delta wing are designed for the supersonic ride. The Rolls-Royce engine reverser buckets help to control jet exhaust flow, reduce noise and slow the aircraft on landing.

Three ministries of the British government fielded teams in 1971 to study the environmental effects of Concorde on humans, livestock and buildings. Damage claims were reviewed by a commission. One claim of $297.60 was paid to a farmer who complained that a cow had stampeded upon hearing Concorde and aborted her calf.

The Concorde's sonic boom can best be described in personal terms. *Aviation Week* wrote in 1971:

It generally is described as a double-roll of distant thunder, although some claim it is a sharper sound, such as a double-clap.

"Weather is a factor in that on a clear day the sound tends to sharpness, and a cloudy day will disperse the noise. It is generally agreed that the noise does not sound like a boom, but more of a rumble. Some persons who have put in claims, however, describe it as an unexpected explosion."

Some other interesting claims:
• The government as of 1971 paid out more than $7,000 in claims for Concorde damages.
• A Scotsman complained that "the heavens opened up and the earth shook when Concorde flew over." He asked for the equivalent of 67 cents for the price of a tube of glue to reseal two tiles to the ceiling of his abode.
• A number of complaints of excessive sonic boom were received even as the Concorde remained in its hangar at the RAF-Fairford test base. Noise was attributed in these cases to garage and car doors slamming, the rumble of heavily laden trucks and distant thunder.

The primary benefit of Concorde – its speed – provides a secondary favor for its passengers. The transatlantic trip on the Concorde transforms a long-haul into a medium-haul trip, and jet lag is virtually unknown. Concorde travellers also can live the unique experience of departing London or Paris at one hour and arriving at a North American destination at an earlier local time.

The first supersonic airliner to fly was the Tupolev Tu-144. The aircraft has been referred to as Concordski, because of similarities to the Concorde. The most obvious design point of agreement is the droop-nose capability. On approach, when the aircraft is nose high, the nose and visor are hinged downward, affording an improved forward view from the cockpit.

The Tu-144 made an appearance in model form at the Paris air show in 1965. Its first flight was recorded in 1968, and the first supersonic flight in 1969. During the lengthy flight test period in the early 1970s, the aircraft underwent a major redesign. The ogival, or pointed, wing of the original design was replaced by a delta wing. The sweepback angle is approximately 76 degrees as it peels away from the fuselage. The Tu-144 was powered by Kuznetsov NK-144 turbofan engines.

The redesigned Soviet SST was exhibited at the Paris air show in 1973. The exhibition aircraft tragically crashed during a performance, and the cause of the crash has not been revealed. The Tu-144, however, entered service in 1975 as a cargo carrier. Passenger service began in 1977 between Moscow and Alma Ata. The services later were suspended, but were started again for a period in the 1980s before final shutdown.

Above: Concorde has an initial rate of climb of 5,000 feet per minute (25.2 meters per second). It usually reaches Mach 1 at 32,000 feet (9,500 meters) and Mach 2 at 50,000 feet (16,600 meters) about 40 minutes after takeoff.

Left: Concorde taking off, leaving a hot trail. Engine manufacturers Rolls-Royce and Snecma joined to develop the Olympus engine. Four 593 engines, each equipped with an afterburner, power the Concorde.

Opposite: Beauty flows from the angular design. Passengers describe the supersonic ride as exhilarating. For the transatlantic blue-ribbon passengers, Concorde wipes out jet lag.

Left: The A300 was the first wide-body aircraft with a two-man crew.

Below: In its day the economics of the 269-passenger A300B2 couldn't be matched by trijet wide-bodies. The A300s may all look alike from a distance, but they are incredibly different in performance. Later versions increased the range. Models such as the B2-300 were designed more durably for multi-stop operations.

SHORT TO MEDIUM RANGE
The Next Generation

Opposite: The 727 was designed for high-frequency operations in the United States. Its short-field performance gave it wide accessibility. Its three Pratt & Whitney JT8Ds produced ample power.

Below: A 727-200, the most popular 727. Re-engining and hush-kitting programs are under way that could permit Boeing 727 operations into the next century.

To many around the world, the Boeing 727 represents the jet transport. It can be seen at most airports in all parts of the globe. Approximately 1,600 727s are flying, accounting for nearly a fifth of the total world fleet. The long, slim Boeing fuselage, the gracefully swept-back wings, the high T-shaped tail and three-engine arrangement in the aft section are familiar characteristics of this popular transport.

In the United States, the Boeing 727 set the standard in the 1960s for short to medium-length domestic routes and paved the way for other turbofan transports. First the Boeing 727 had to prove it could perform on short runways and hold its own economically against turboprop transports, specifically the Lockheed Electra. Boeing design concentrated on development of a high-performance wing and strong landing gear.

> "We built a jet airplane to get in and out of a 5,000-ft. field. No one believed it could be done."
> – **Joseph Sutter, executive vice president (retired), Boeing Commercial Airplanes, on the Boeing 727**

The 727 became the launching platform for a new powerplant, the Pratt & Whitney JT8D, which has become the most popular turbofan jet engine in the transport fleet. For example, Pratt has sold 12,500 JT8Ds as of December, 1987. Powered by the three JT8Ds at 14,000 lb. take-off thrust each and raised to new capability with a high-performance wing, the Boeing 727 carried out the short-field demonstration at the 5,000-foot-long Runway 4 at New York's La Guardia Airport without a problem.

Among its characteristics, the 727 wing has:

• A 32-degree swept wing at the quarter-chord.

• Inboard ailerons for high speeds and outboard ailerons for low speeds.

• Triple-slotted trailing-edge flaps.

• Four leading-edge slats.

• Three Krueger leading-edge flaps in the inboard section of the wing.

• Seven spoilers on each wing, five for flight and two for ground use.

"It was a lucky airplane," Joseph F. Sutter now a retired Boeing executive, said. He cites several examples.

Left: The three-man crew and three-jet operations represented 1960s economics, but the 727 has remained useful for its owners and operators into the airline deregulation era.

When the Boeing 727 was landing at La Guardia for an evaluation at the runway by noise-sensitive politicians, crosswinds pushed the jet noise away from where they were standing. "With the crosswinds that day the politicians thought the airplane didn't make any noise. Of course a little bit of luck favored us on that day."

The 727 was fortunate in its timing as well. John E. Steiner, chief engineer on the program, accords two lives to the 727. The first started in 1964 with initial deliveries of the 100 Series transport, which accommodates up to 130 passengers. The second began in 1970. It was about this time that it became clear to airline managers that wide-body aircraft were too large for the high-frequency services desired on the developing route networks in the U.S. The stretched version, the 727-200, accommodating up to 189 passengers, became the transport of choice for high-frequency flying.

Eastern launched the 727-100 into commercial service in February, 1964. The stretched 727-200 was certificated in November, 1967. Its peak year, 1968, saw 160 727s delivered, but the sliding economy caused a drop in deliveries in the early 1970s to a steady but lower number. The early years of airline deregulation in the United States influenced sales once again, and 727 deliveries climbed to 118, 136 and 131 a year in 1978, 1979 and 1980 respectively. In all, 571 727-100s were produced, and 1,260 of the stretched 200 Series were built. In 1984 Boeing made its last eight deliveries of the stretched version.

Above: An S-shaped duct carries air from the middle engine inlet to the engine itself, which exhausts through the tail cone.

Opposite: The tri-jet Boeing 727 has established itself firmly in aviation's history. With 1,600 or more 727s flying in most parts of the globe, it accounted for nearly a quarter of the world fleet.

The 727 was the first commercial airplane to surpass the 1,000 sales mark for civil use.

Alaskan Airlines exploits the vertical stabilizer to herald Alaskan Eskimo.

The Twin Alternative

Opposite — top: DC-9s are powered by Pratt and Whitney JT8D engines, the world's most successful commercial turbofan.

Opposite — bottom: DC-9-10 has rear-mounted engines, just like the Caravelle. A reason for positioning was to reduce noise in most of the passenger cabin.

DC-9-10, the first version of this aircraft.

The Douglas Aircraft Co. joined the competition for a short-haul airplane. Douglas designers had their eye on an exclusively short-haul transport that would be small and economical in comparison with the Boeing 727. The first version, the DC-9-10, nestled firmly in the developing market in 1966.

Captain Alfred Born of Swissair, pilot of early Douglas DC-9 models, compared the DC-9-10s handling to that of a fighter aircraft.

In the DC-9 airline management found economy with two crew members and two engines. The DC-9 set a precedent as America's first twin-engine aircraft with rear mounts.

The DC-9 Series showed remarkable application to airline needs around the world. The DC-9-10, for instance, served as launch aircraft for Chicago-based Midway Airlines in 1979 in the early days of U.S. airline deregulation. Air Canada deployed DC-9s throughout its system of short- and medium-haul routes. Among the later versions, the MD-80s are in use in China and are being co-produced by McDonnell Douglas and the Chinese.

Delta Air Lines, launch customer for the DC-9-10, enjoyed a smooth introduction of the aircraft in early 1966. Inside of a year after training started in October, 1965, Delta was operating the aircraft at an average of almost nine hours a day and had few break-in problems. The average DC-9-10 route was 269 statute miles long, a 42-minute flight on average.

The 115-seat DC-9-30 was the most popular of the six DC-9 versions. More than 500 Series 30s were sold. Its gross weight of 109,000 lb is half that of the Boeing 727. Its range of 1,800 miles, with maximum payload, classifies it as short-haul. Douglas stretched the DC-9-10/20s, each 104 feet, 4.75 inches long, to 119 feet 3.5 inches long for the Series 30 version.

Douglas' latest version, the MD-80 Series, has taken on almost a life of its own. First known as the DC-9-80 with a 147-foot, 10-in. long fuselage, the derivative is now known as the MD-80 Series, and has varying capacities. The McDonnell Douglas MD-80 was a test bed for several proposed high-technology propfan engines, but the program has lacked airline interest.

Left: Nose of a DC-9-50.

Opposite: The short-range, low-capacity models soon gave way to stretched versions. The MD-80 Series owes its origin to the DC-9-80.

Below: The twin-engine, two crew DC-9 was a desirable aircraft for high-frequency, low-cost trips under USA airline deregulation.

Top: 737-400.

Center: The CFM56 engine powers the more recent versions of the 737 such as the -300, -400 and -500 series.

Right: 737-500.

TODAY'S PRODUCTION AIRCRAFT

The Search for Efficiency
Boeing's 757 and 767

The development of the Boeing 757 and 767 was influenced strongly by the availability of more efficient turbofan engines and other new technologies. In a sense, the fuel crises of the 1970s set the stage for the new transports.

Conventional in appearance, look-alike cousins of earlier Boeing jet transports, the 757 and 767 twins have taken advantage of new-generation technologies offering operational economies and new efficiencies. The differences from earlier Boeing models are unseen by the public, but are plentiful in the cockpit. Automation and the use of new techniques for navigation and flight control have placed the crew members in the roles of monitors of systems, rather than hands-on fliers. The 757 and 767 are now being used in a variety of roles that will keep them in the skies well beyond the turn of the century.

Designers at Boeing in the early 1970s had in mind a single aircraft that would be able to exploit the developing new technologies. In addition to more efficient turbofans, there were improved aluminum alloys; materials made from composites, also known as advanced structures; the latest in avionics, including displays; and more efficient wings. The technologies could not have become available at a better time. In late 1972, a fuel boycott threatened scarcity, and in 1973 prices soared. Inside of two years, fuel prices doubled. While fuel in the pre-crisis days had equalled other costs, such as crew costs and maintenance, it became the biggest expense by far.

A sick customer results in a sick airplane manufacturing industry, whatever the cause may be.
– John E. Steiner, Boeing Commercial Airplanes (retired), in Jet Aviation Development, A Company Perspective

Under exploration at that time was a twin turbofan-powered 727. But the fuel crisis and sluggish traffic weakened the airlines financially. It took until 1978, a boom year for airlines, for Boeing to launch the 767 and shortly thereafter the 757.

The 757 adopted the same fuselage shape and size originated by the 707 and used by the 727 and the 737. The 757 is easily distinguished from the 767 by its nose, which is uncharacteristic of bullet-nosed Boeing transports and more like those produced by Lockheed. The flattened out, broader-based nose resulted from wedging the 767 cockpit in the smaller 757 frame.

Right: 767 and 757 cockpits are the same technically. Pilots are qualified to fly in both aircraft. One difference is the height from the ground. The 767 cockpit is a few feet higher.

Opposite: The 757-200 retains the same fuselage cross-section as the 707, 727 and 737. This versatile aircraft is flown on a wide variety of routes. The twinjet is used to serve routes as far as 4,000 miles (6,500 km) and as close as 75 miles (120 km) on shuttle flights.

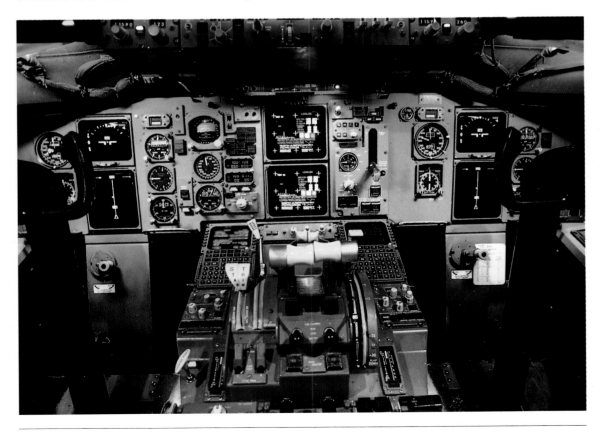

Left and below, far left: Airlines are offered a choice of powerplant for the Boeing 757-200. The Rolls-Royce RB211-535E4 engine was Monarch's choice. It has compiled an enviable record for performance. Delta chose the Pratt & Whitney PW2037 engine. It offers 38,200 lb. takeoff thrust.

Below: The original design for the 757 provided for a T tail configuration, but as a result of wind tunnel testing it was decided to mount it on the fuselage.

Boeing at first presented the 757 as a 150-seat transport. Frank Borman, then head of Eastern Airlines, persuaded Boeing to increase the passenger capacity. Eastern had acquired the high-capacity Airbus and was stuck with operating low-capacity 727-100s, and it needed something in between. The 757 accommodates 186 passengers in typical U.S. seating or as many as 230 passengers in an all-tourist configuration.

When Borman thinks of the 757, he thinks of the great reliability and its extremely low operating cost. Eastern recorded lower costs on the basis of seat miles (the number of seats multiplied by miles traveled) than the costs of operating a 747 on transcontinental-length routes.

The Boeing 757 was a slow starter for Boeing Commercial Airplane Co. A slump in sales followed initial orders from Eastern Airlines and British Airways. Delta Air Lines caused a spike in the 757 order book in the early 1980s. Slowly other carriers, such as Northwest and American, joined the 757 fan club. Fuel efficiency and flexible capacity, 186 passengers to an all-tourist configuration of 230 passengers, made the twin-engine transport an economic replacement for smaller-capacity airlines on routes of growing traffic demand.

The Versatile 767

Boeing's 767, designed to work in tandem with the 757, is becoming even more versatile. The transport originated in the 200 series. An extended range version became available in the ER series, and a 300 stretched version was followed by three new configurations to meet a variety of long-range roles.

Originally the 767-200 was designed to wedge itself between the two basic body types – wide-body and narrow-body transports. A new cabin size of 15 feet 6 inches was adopted, which compares with the 747's 20-foot 1-inch interior. Amenities of the wide-body were retained in the 767: specifically the twin-aisle cabin, an 80-inch-high ceiling, and cabin stowage space nearing two cubic feet per passenger.

The 767 had tentative acceptance from passengers when United placed it in service in late 1982. Frequent fliers compared it unfavorably with the wider wide-bodies. Dispatch reliability, representing the readiness of the aircraft to operate, was at a low 90% in initial months. The low reliability was due to disruptions of digital computer systems when cockpit crews switched from ground power to the aircraft's own electrical generating system. Computers went off line, lost stored information and had to be reset carefully to avoid disturbing delicate system monitors. The resetting was time-consuming and required precise work that led to delays. Eventually, mechanics and crews grew more adept at handling problems.

The 767 provided some of commercial aviation's greatest dramas in the 1980s. An Air Canada 767, short of fuel, had a double flameout at an altitude of 40,000 feet and glided powerless for 15 minutes to a no-flaps emergency landing in twilight on July 23, 1983 at a former Canadian Armed Forces air base at Gimli, Manitoba. At the time, Boeing officials said the 767 established a record for a glide in terms of the aircraft's weight and altitude. In other words, it was the biggest and highest-flying powerless aircraft ever brought down to earth safely. The Air Canada crew was disciplined for a fueling error, which the Canadian flag carrier blamed on incorrect converting of metric weight and volume information. Air Canada 767s were the first to use metric measurements.

Fated for the fueling error, the 767 incident had its own set of fortunate circumstances. The Air Canada first officer, Maurice Quintal, served as a Canadian officer at the Gimli base and directed the pilot, Captain R. O. Pearson, to the landing site. Pearson was a glider pilot. On final approach without engine power, Pearson performed a maneuver called a sideslip to bleed off altitude and touched down "on the numbers" of Runway 32L.

On August 19, 1983, a United Airlines 767, powered by Pratt & Whitney JT9D-7R4 engines, glided powerless for several thousand feet. The crew first shut down one engine because of suspected overheating. Then the second engine recorded similar overheating and was shut down.

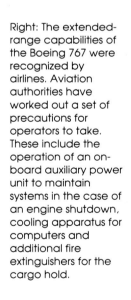

Opposite: The 767 was the first Boeing airplane to replace aluminium with graphite and Kevlar composites, which are lighter, stronger materials. The weight savings contribute to overall operating efficiency.

Right: The extended-range capabilities of the Boeing 767 were recognized by airlines. Aviation authorities have worked out a set of precautions for operators to take. These include the operation of an on-board auxiliary power unit to maintain systems in the case of an engine shutdown, cooling apparatus for computers and additional fire extinguishers for the cargo hold.

Both were restarted after a glide for a powered landing at Denver Stapleton International Airport. Among the resulting modifications by Pratt & Whitney was one for improved engine performance at low idle.

The 767 has led the way for twin-engine transports in long-range overwater and other extended-range operations. The aircraft operate under specific rules. They are equipped with an auxiliary on-board powerplant to keep flight systems operating in case of a shutdown of one engine. There are special cooling devices for computers and more fire extinguishers than typical for added protection. Engine-monitoring devices in the cockpit are closely watched for indications of potential problems. At first sign of an engine irregularity, for instance, the engine may be shut down as the aircraft heads for its alternate airport. Because of such precautions and a special rating for extended range crews, the authorities have permitted these twin-engine aircraft to operate on tracks that will take them up to 120 minutes' flying time on one engine from a suitable airport.

El Al Israel Airlines launched the extended-range flights in late 1984 from Montreal to Tel Aviv, followed closely by Air Canada and Trans World Airlines on other transatlantic flights. Trans World has accounted for most of the 767 extended range operations, building from a rate of approximately 150 flights each month in 1986. The airline has had the most diversions to alternate airports under the special procedures. These included two diversions in 1985. In 1988, two Trans World 767s on transatlantic flights were diverted to alternate airports within a five-day period. In each case, engine problems were indicated. The pilots shut down the troublesome engine and flew on a single engine to the alternate airport.

Boeing's 767-300 entered commercial service in October, 1986, and an extended-range version of the -300 began flying commercially in February,

An important element among the precautions required for extended-range flight is experience of a flight crew with a particular engine and airframe combination.

1988. The ER version has a range of more than 6,000 nautical miles.

The -200 and the -300 and each of the ER versions comprise the four available 767 models. A freighter is the fifth model, a development of the 1990s.

More than any other aircraft, the Boeing 767 deserves recognition as the pioneer of extended-range flying.

Right: More than a hundred microprocessors and several computers allow the 767 to fly automatically, from takeoff and climb to landing.

The 767 holds the distance record for commercial twinjet aircraft with a 9,253 mile (14,890 kilometer) nonstop flight from Seattle, Washington, to Nairobi, Kenya, on June 12, 1990, in 17 hours, 51 minutes.

British Aerospace 146

A turbine-powered transport developed for commuter airlines having outstanding success for its low-noise characteristic is the British Aerospace BAe 146. The high-wing, four-engine transport now comes in three versions, a 100, 200 and 300, seating 69, 85 or 100, respectively.

That the BAe 146 is powered by four engines is an anomaly among contemporary jet transports. The Avco Lycoming ALF502R powerplants, which also serve helicopters, develop nearly 7,000 lb. thrust each. They have fewer parts than higher thrust engines, require less maintenance and have a comparatively modest fuel burn. The 146 was not, however, a competitive aircraft in the harsh economics of the post-deregulation era. USAir, which inherited a fleet of 146s from its merger with Pacific Southwest Airlines, parked that fleet in the Mojave Desert in 1992 after the carrier was humbled in competition with Southwest Airlines in the California market.

British Aerospace turned to the 146 as a bed for its next regional offering, the RJ70.

The British firm has joined with Taiwan Aerospace and Avro International to build an array of 70-115-seat Regional Transports. The aircraft are now called the RJ70, RJ85, RJ100 and RJ115, succeeding thte BAe 146 series aircraft. The 120-seat RJX has twin engines.

Ironically, the 146, quiet to the people down below, is mechanically noisy to the passengers. The chief reason for the comparatively higher level of cabin noise is the high wing. Flap action takes place above a passenger's head, not well below it as in low-wing transports. Movement of flaps, letting down of the landing gear and retracting the gear are noticeable but not disturbing events in the 146.

A low-frequency rumbling noise at cruise speed, heard as the 146 was introduced, was toned down in a modification. The source of the rumbling noise was a slot opening in the fuselage in which the flap moves. Between 250 and 290 kt, the opening, combined with air rushing past, emitted the noise. Improved seals have solved that problem.

Another noise problem is one that continues to be endured. A whistle is heard when flaps are retracted after takeoff and extended in preparation for landing. It rises in pitch and volume between 0 and 18 degrees of flap extension.

Air Wisconsin achieved a 99.2% dispatch reliability in the first few months of operating the 146. This reliability is defined by World Airline Technical Operations Glossary (WATOG), and represents superior performance.

But these are minor points for an aircraft that can fly most anywhere in the world because of its low-noise characteristics. British Aerospace has sold more than 140. The aircraft has been honored by the British Royal Family and is operated in the Queen's Flight of the Royal Air Force.

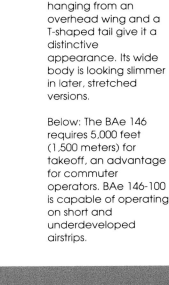

Opposite: The Quiet One. The BAe 146 can fly almost anywhere, because its four Avco Lycoming powerplants are among the least noisy engines in today's fleet. The four engines hanging from an overhead wing and a T-shaped tail give it a distinctive appearance. Its wide body is looking slimmer in later, stretched versions.

Below: The BAe 146 requires 5,000 feet (1,500 meters) for takeoff, an advantage for commuter operators. BAe 146-100 is capable of operating on short and underdeveloped airstrips.

Left: Four Avco Lycoming turbofans power the British Aerospace BAe 146. Mounted on a high wing, the engines are well clear of the ground.

Opposite: The BAe 146 is the only jet allowed to land at London's City airport.

Below: The original BAe 146-100 has been stretched twice to accommodate more passengers. The new, larger models are designated -200 and -300. The Royal Air Force acquired the BAe 146 for the Queen's Flight, a tribute to the aircraft.

Right: Tupolev Tu-154M is normally laid out to accommodate 158 to 164 passengers in a single-class layout. It has a range of about 3,000 miles (4,800 kilometers) with a full passenger load.

Middle: The Tupolev Tu-134A twinjet has a slightly longer fuselage and nose radar to distinguish it from the TU-134. Both models have proved to be among the most popular of Russian aircraft for export.

Bottom: The first flight of the Tupolev Tu-154M was made on 4 October, 1968.

Below: An A320-200 of Air France. The A320-100 was offered originally, but was replaced from late 1988 by the -200, which has wingtip fences, a wing center-section fuel tank and a higher takeoff weight.

Opposite: British Airways was among the first airlines to order the Airbus A320. Its high technology and fuel efficiency make the aircraft an attractive replacement for the Boeing 727. The pilot flies the aircraft through a computer, and the flight control surfaces, such as the elevators and ailerons, are activated electronically. This fly-by-wire technology replaces hydraulic systems, which have been the mainstay of flight control systems for decades.

NEW TRANSPORTS
The Advanced A320

The fly-by-wire technology in the flight control system of the Airbus A320 qualifies that aircraft for the title of the world's most advanced large transport. Flight control surfaces – elevators, ailerons and spoilers – are activated by electronic impulses. Hand control of the aircraft in the cockpit comes through sidestick controllers – joysticks – not the usual control yoke, a break from tradition.

Airbus' A320 is flown through a computer, much like the other contemporary transports. The software has numerous protective features for the operation of the aircraft, based on the aircraft's angle of attack. For instance, the software precludes the pilot from applying "too much stick" for maximum climbout that, under similar circumstances in another aircraft, could lead to a critical angle and pitch for flight – and a wing stall. A primary advantage in being able to apply maximum, no-stall stick is speedy and appropriate pilot response to a possible encounter with wind shear, a sudden change of wind speed and direction that can destroy lift for an aircraft.

Pilot reports on the fly-by-wire system and the sidestick controllers have been favorable. Generally, pilots have said that aircraft handling is improved, crew workloads are reduced, and the systems offer increasing protection against stalls and wind shear. The software also provides the aircraft with a special sensitivity to wind gusts and turbulence. Flight control surfaces are deployed to reduce the effects of turbulence on the aircraft, resulting in a smoother ride for passengers.

The cockpit area is wide and one can easily slip into the pilot seats without the contortions required in some transport aircraft. This ease of entry is aided by the absence of a control column. The left armrest in the pilot's seat is adjustable so that the left arm can rest easily and facilitate wrist movement in grasping the sidestick controller.
– Pilot report on the A320 by David M. North, Aviation Week & Space Technology

Airbus demonstrated the fly-by-wire system first on an A300 test bed at the Farnborough air show in 1986 and again in an A320 at the Paris air show in 1987. The system greatly improves low-speed flying qualities. At Paris, the A320 executed a low-altitude, low-speed pass as if it were in landing configuration, with landing gear and flaps down. The crew initiated a go-around during the performance. The aircraft's nose pitched up in spite of the low-energy state of the aircraft. Engine power increased, and the aircraft recovered.

The A320 entered commercial service in Europe in 1988. It is the first Airbus product that has won substantial orders from North America. Airbus operators included the defunct Pan American World Airways, Continental Airlines, Northwest Airlines and United Airlines are acquiring the A320, a major breakthrough for the consortium. Additionally, Air Canada operates the A320. Taking advantage of the 3,000-nautical-mile range with 150 passengers and baggage, Air Canada planned to operate the A320 coast-to-coast on trans-Canada routes and from Canada to South America.

Once in the left seat, I found that the A320 pilot seat was comfortable and the visibility excellent, including the ability to see the left wingtip without much effort. The sidestick controller was easy to grasp with the left arm resting on the extended arm rest. Five minutes at the most was required to become accustomed to the sidestick controller, which differs from the control column installed in most transports and the stick found in fighters."
– Pilot report on the A320 by David M. North, Aviation Week & Space Technology

Airbus has built a stretched version of the A320, the 186-seat A321, which entered production at the Airbus Industrie factory at Hamburg, Germany, in 1992. The consortium has planned a shortened version, the 127-seat A319, which would be a replacement for early versions of the Boeing 737.

In its original configuration of 12 first class passengers and 138 in coach, the A320 served as a replacement for the Boeing 727. The nearest competitors the -400, and the MD-80 Series aircraft.

Above: CFM International CFM56 engine showing bird eye design to scare birds off at takeoff. The CFM56 powerplant is jointly manufactured by General Electric and Snecma of France.

Opposite: Typical of contemporary commercial aircraft, the A320 is flown by a two-person cockpit crew. The Airbus cockpit is the newest adaptation of the pilot's workroom. Its various displays are uncluttered even to such instruments as the horizon indicator. Airbus broke tradition by eliminating the control yoke and replacing it with a sidestick controller.

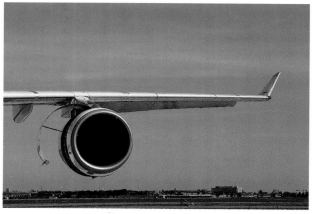

The A340 shares its design with the A330. The only visible difference is the deletion of the A340's two outer engines. The wings have upward-pointing winglets similar to the 747-400's.

The A340 cockpit corresponds almost exactly that of an A320. The most important differences are the advanced flight management and navigation systems and controls for four engines instead of two.

Airbus Alternatives

Both Lufthansa and Air France began operating the four-engine A340 across the Atlantic in 1993. The operations were the culmination of many years of labor, based on an Airbus Industrie concept more than a decade old.

The A340 was conceived as a long-range aircraft seating between 260 and 295 passengers. An aircraft with high dispatch reliability for the long-haul routes was envisioned. The 200 series now in operation is configured for seating approximately 260 passengers. The cargo load is about 52,000 lb.; the range nears 76,000 nautical miles. The 300 series aircraft seats close to 300 passengers and takes minor penalties for range and cargo load, compared to the 200.

The A340 flight deck is similar to that in the A320, which brings a certain welcome familiarity to Airbus cockpits.

A sister to the A340s is the A330 twin-engine transport, which has had pre-launch orders, as of February, 1993, totaling 144 aircraft. These will be operated in a typical configuration of 305 coach seats and 30 first class seats. The range is up to 5,000 nautical miles. maximum gross takeoff weight is more than a half-million pounds.

A340 production at Airbus Industrie in Toulouse, France. The flight test program for certification took around 2,000 hours of flying time and six aircraft.

Boeing Family

In late 1989, the Seattle-based manufacturer offered a new wide-body, twin-engine aircraft, the Boeing 777, an alternative transport sized between the 767-300 and the 747-400. An order from United Airlines launched the program in October, 1990, and orders have been plentiful since then.

The 777 will be the largest twin-powered transport. An all-economy, 10-abreast seating configuration will provide for 440 passengers. In three-class configuration with seats in nine-abreast rows, the capacity still will be more than 300 passengers.

Boeing's latest offering is scheduled for its first flight in 1994, and in-service operation with an airline the next year.

The first version, the Boeing 777-200, will replace aging tri-jet transports currently operating in transcontinental markets in the United States, Europe to the Middle East, and routes in Asia. The range will be approximately 4,000 nautical miles. It will offer between 20 and 25 percent more capacity than the aircraft it replaces. By December, 1995, the second version, having increased engine thrust and greater takeoff weight, will be capable of flying 6,000 nautical miles. This intercontinental aircraft will be flying between the West Coast of North America and Europe, and from the Asian mainland to Australia.

Boeing expects the 777 will grow and generate a family of aircraft of its own. By the late 1990s, a stretch version of the regional aircraft is under consideration. One that could fly 8,000 nautical miles is likely to be in Boeing's future.

The Boeing 777 will be the first "paperless" commercial airplane, one to be defined entirely on a computer, not a drawing board. Each part is modeled on the computer in three dimensions, placed in the computer-projected aircraft and displayed in three dimensions.

United and other launch airlines are participating is some of the 217 design/build teams for the 777. These teams are working to resolve conflicts before aircraft assembly.

Two other projects are under study at Boeing:
• The Super Jumbo Jet, a four-engine transport successor to the Boeing 747, with seating capacities of 650 passengers and higher. The aircraft is also known as the Very Large Commercial Transport (VLCT), and is under study by Boeing in concert with European manufacturers.
• The Next-Generation Supersonic Transport (SST), projected to operate at around Mach 2.2–Mach 3 and carrying 300 passengers on intercontinental ranges.

Newest Aircraft

Sales of the various types of the Boeing 737 – the 300, 400 and 500 – have broken records and certified the 737 as the world's most popular turbofan-powered transport with more than 3,000 aircraft ordered. The fact that there are so many 737s, featuring a high degree of commonalty of parts and systems, assures the aircraft of longevity. Commonality of parts and systems reduces the inventories and training costs of airlines, another reason to continue buying and using 737s. The former Piedmont Airlines, now merged with USAir, launched the 737-400, and began using that aircraft in revenue flights in September, 1988. The aircraft is a part of the USAir fleet currently.

The latest 737 is the 737-500, a smaller version more like the original 737-200. The aircraft is configured for a mixed-class seating of 108 passengers and an all coach seating of 132. Rollout occurred in 1989, and deliveries began in 1990.

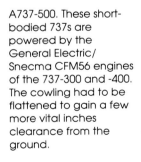

A737-500. These short-bodied 737s are powered by the General Electric/Snecma CFM56 engines of the 737-300 and -400. The cowling had to be flattened to gain a few more vital inches clearance from the ground.

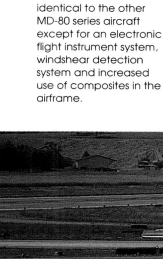

Opposite: The MD-87 is a short-fuselage version of the MD-80. It is similar in length to the DC-9-50 but has the more powerful Pratt & Whitney JT8D-217 engines of the MD-82 and 83.

Below: The MD-88 is identical to the other MD-80 series aircraft except for an electronic flight instrument system, windshear detection system and increased use of composites in the airframe.

The Douglas Entries

The latest Douglas entry is the MD-90, a derivative of and a distant cousin to the DC-9. It is about five feet longer than the standard MD-80 series aircraft, its immediate predecessor.

Records achieved in initial flights in 1993 promise a future of low-noise operations. Douglas officials predicted the MD-80 will be 60 percent quieter than the MD-82 on takeoff. This is no mean feat when one considers that the MD-82, in compliance with Stage 3 noise regulations, is already among the quietest transports in the air.

The first MD-90 is powered by International Aero Engines (IAE) V2500 turbofans. The aircraft obtained additional noise reduction from modifications of the Rohr-produced engine nacelles. A 15-inch nacelle inlet extension provides additional acoustic benefits in the front end; acoustic treatments to the jet pipe and tail cone also result in noise reduction.

Douglas' MD-90 is competing against the Airbus A320 and the Boeing 737-400. The manufacturer anticipates a whole series of MD-90 models, just as it produced a variety of aircraft under MD-80 nomenclature. The first MD-90 will be a -30. Passenger capacity will range from 153 in a two-class configuration to 172. The range with 153 passengers is projected to be approximately 2,200 nautical miles.

In the late 1980s, Douglas flew an MD-80 demonstrator aircraft powered by one ultra-high bypass propfan engine, the General Electric GE36 Unducted Fan (UDF) and a conventional turbofan. Douglas also demonstrated the Pratt & Whitney PW-Allison Model 578-DX ultra-high bypass engine. Huge fuel efficiencies were promised, but airline interest was never really strong, in part because of concerns over passenger reactions to noise.

Delta Air Lines is a customer for the MD-90. An earlier version, the 143-passenger MD-88, was virtually developed for Delta. It was modeled after an early 1980s design by Delta's own engineering department that the carrier called the Delta 3.

The MD-80 series borrowed from military technology the head-up displays in the cockpit.

Bottom: An MD-87 of CTA of Switzerland. The Geneva-based charter company merged with its sister company Balair.

The MD-11 and Stretch

The aircraft closely resembles the DC-10. The preliminary differences lie in range, passenger load and in the avionics suite.

Cockpit development was the work of pilots from 37 airlines. Automated systems can be overruled by the pilot.

The MD-11 has overcome initial performance problems. Fuel burn fell short of expectations. The Douglas goal from the beginning has been to carry a full load – 267 passengers – for a 7,000-nautical-mile trip.

Douglas has considered two stretch versions of the MD-11. The manufacturer has explored a variety of options for an MD-12, a double-decker transport with huge passenger capacity. But financial reverses at the company and in general have slowed development of programs.

Aerodynamic improvements over the DC-10 include winglets and redesigned wing trailing edge, a smaller horizontal tail with integral fuel tanks and an extended tail cone.

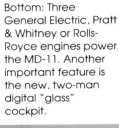

Bottom: Three General Electric, Pratt & Whitney or Rolls-Royce engines power the MD-11. Another important feature is the new, two-man digital "glass" cockpit.

Vapors from jet aircraft criss-cross the sky over New York. The three major airports of the area handle more than a million aircraft operations per year. The airways are the world's busiest

Opposite: Zurich airport's control tower and radar.

Airline passengers grew to more than one billion in the year 1988. The world-wide industry group, the International Air Transport Association, forecasts that passengers will double by the turn of the century.

Below: New control tower at Geneva's Cointrin airport. An A310 in the foreground. The Airbus is flown from Geneva to other major cities in Europe, and to the Middle East and Africa.

Each day brings its own events, its own twists and developments that stamp its character – fundamental conditions such as the weather. The forecast is an essential part of the pre-flight briefing. As the crew gathers for this briefing, of primary concern is the amount of fuel to be loaded on the aircraft, cruising speed given the expected head-winds or tailwinds, temperature at altitude, and the number of passengers, each of which affects the fuel consumption rate.

A typical summertime flight on a Swissair Boeing 747-300 from Geneva, Switzerland, to New York, will consume 96,000 kilograms of fuel (212,160 lb.) in 7 hours and 55 minutes against fairly strong westerly winds. The return trip, flying with the predominant westerlies, will consume 75,000 kilograms of fuel (165,750 lb.).

Alternate airports are selected along any airline route in case there is a need to land quickly. Pilots say the forecast weather is generally accurate, but not always as predicted. Many aircraft are equipped with radars that display on cockpit screens the weather patterns 30 to 300 nautical miles ahead in color codes that define storm intensity. There are turbulence plots available on new-technology transports to pinpoint rough air miles ahead.

"A max gross takeoff in a whale (a Boeing 747) is something that everyone should experience once in his career."
– Captain Patrick Clyne, Northwest Airlines

Aircraft dispatch has been an important part of flying since the beginning of commercial aviation. Wilbur Wright, standing by as his brother Orville took the controls of a 16-horsepower engine, served as the first dispatcher. Wilbur also piloted the biplane, flying the fourth flight on that momentous day and covering 852 feet in 59 seconds.

Each day, dispatchers are on duty to prepare flight crews and attend to the flight. Many dispatchers at John F. Kennedy International Airport in New York are located in offices overlooking the vast aprons where aircraft come and go. They talk via radio to incoming and outgoing aircraft, giving the crews updated weather information. Swissair's own high-frequency (HF) radio at JFK can reach aircraft an hour's distance over the Atlantic Ocean from New York. With the help of Aeronautical Radio (ARINC), a company owned by the airlines, a Swissair call can be "patched through" to an aircraft at mid-ocean at 30-degrees West Long. Facilities in Europe can reach the aircraft from the European side of the Atlantic Ocean.

Opposite and this page: Air traffic control is a highly automated process. Towers at airports are the headquarters for local control. Responsibilities are divided among controllers who separate traffic on the ground and the aircraft on approach and those taking off.

Passenger growth has placed great stress on airports around the world. The use of curfews to halt air operations at certain times of day has spread. Few new airports are being built, but many are expanding.

London Heathrow airport's radar tower.

To the dispatcher falls the job of providing the pre-flight briefing to the cockpit crew. Today, the dispatcher uses a computer to work out the best possible route from the standpoint of winds, the most economical cruise altitude, and estimated use of fuel. To the dispatcher, flight today "is all paperwork and computers."

As always, the pilot in command has the last word on the amount of fuel to be loaded on the aircraft. Such decisions are basic to the flight of which he is in charge. The flight track is selected by the pilot or, in many cases, simply approved by him. A flight plan is filed, notifying air traffic control authorities of the aircraft's path through national and international airspace. The crew records its current airport position in longitude and latitude on the inertial navigation system in the cockpit. This step lets the INS know where the aircraft is located on a global map. The crew then is able to select up to nine points along the route, called waypoints, where reports are made to air traffic control and fuel checks are made. Long trips require the crewmembers, while they are in the air, to select waypoints on the track ahead. A Chicago-to-Seoul flight for instance, will involve as many as 30 waypoints. They are nowhere special, just reporting points fixed on a map. The flight plan contains estimates that indicate the time of flight between waypoints and the fuel use.

The 5,978-nautical-mile trip between Chicago and Seoul, South Korea, in a Northwest Airlines Boeing 747-200 will cover 30 waypoints over the United States, Canada, the northern Pacific and Japan, straddling Russia's airspace. On such a flight, lasting more than 14 hours, there are two cockpit crews aboard. One is an "augmenting" crew that relieves the start-off crew.

Once the crew is given final information on the aircraft's weight, the vital numbers for takeoff can be calculated. The first is a decision speed. This is known as V_1 – Velocity 1. When this speed is reached, the captain must make a decision to go or not go. The takeoff can be aborted up to the decision speed. The second is the takeoff speed known as V_r – Velocity rotate. At this point, the aircraft actually rotates – the nose pitches upward – and the wings begin to ride on the air.

The United States has undertaken a $14-billion modernization of its air traffic control system. New computers already have been installed at key centers around the nation. The Canadians have developed their answer to traffic growth in a new ATC system. Other nations are following these leads to prepare for the growth spurt in the 1990s. Along the concourses in terminals everywhere, the aircraft dock at gates. Here they are replenished with fuel and receive ordinary maintenance. Electric power is provided. Provisions for the passengers are available. And through jetways, the passengers board.

A 747-400 and its crew preparing for a flight. With this aircraft, 10- to 14-hour flights have become routine.

Top: The nosewheel of a 747 with the ground power cables attached.

An aircraft must obtain clearances from air traffic control for its movements on the ground and in the air. The crew begins a takeoff checklist. The setting for the wing flaps that will improve lift are among the items ticked off. At Northwest Airlines and other airlines, a takeoff card is prepared. Once cleared to go, the pilot has in his hands the throttles for 200,000 lb. of thrust from the four engines. At full throttle the engines roar and the aircraft rolls and accelerates to decision speed, to rotation speed, to liftoff and climbout. By the time the flaps are retracted, the aircraft will be flying at well over 200 mph.

On an actual flight of a Northwest Boeing 747-200 at maximum gross weight of 800,000 lb. between Chicago and Seoul, the numbers were as follows:

- **Decision speed known as V1, 152 kt (281 kph).**

- **Climbout speed V2, 175 kt (324 kph).**

From this point on in the flight, the crew begins a regimen of position checks, communications with ATC authorities and monitoring of instruments, fuel burn and the sky around them.

Flights end with a final flurry of paperwork and checking of fuel use, observing errors in navigation and maintenance problems. Usually on long-range trips, the crew will be called upon to fly another flight in 20 hours. The flights will be part of a 12-day trip for the crewmembers who will pass through as many as a dozen time zones in each flight. Such a disruption of normal life habits has its effects.

"There is an adage that when you leave the house for a 12-day trip, that's as good as you're going to feel for the next two weeks," says Northwest's Captain Patrick Clyne.

One can understand the flight as the basic work of an airline in the many preparations that begin well before the listing of the flight on an airport departure screen. Swissair's catering department at Zurich, one of four operated by the Swiss flag carrier, is the center for the production of 20,000 and more meals each day, more than seven million meals a year. The 250 cooks and chefs will prepare a ton and a half of meats on an average day.

In 1987 the Zurich kitchens of Swissair produced 7,260,000 meals – 30% of these for other airlines. The 250 cooks and chefs prepared one and a half tons of meat each day and served seven tons of caviar in the year. Swissair has defined nearly 300 meal cards or separate menus. Six different menus are available any week in Europe.

The food prepared for the passengers serves to illustrate the basic internationality of airlines. Of these seven million meals, 22,500 were Asian vegetarian, 23,000 were Western vegetarian, 9,500 were Moslem, 1,200 Hindu, 1,800 diabetic and 6,000 meals were for other diets, primarily salt-free.

The commissary at Zurich stores the basic and extraordinary items for the service of passengers. One can find medical kits, toys for children, fine wines and food delicacies, stacks of daily newspapers, magazines, thousands of pillow cushions and blankets, headphones – all the conveniences.

Opposite: Loading of a 747 with meals, cargo and mail before transatlantic flight.

Above: Loading of cargo before flight. Most large aircraft have front and rear cargo loading doors. The procedure is fullly automated and is usually done by two people. Cargo revenues represent up to 20 percent of an airline's yearly sales.

Opposite: A 747-200F cargo aircraft. This type has the ability to open its nose and load freight straight-in and has a maximum lifting capacity of 250,000lb. (113,600 kg.).

Right: The push back tug is attached to the 747. These heavy vehicles weight as much as an eighth of the aircraft they pull or push.

Boeing 747 is pushed off the gate. The Boeing 747-300 series weighs, on average 383,000 lb. (174,000 kg.) when empty. It can lift 833,000 lb. (378,600 kg.). The difference is made up of fuel, provisions, passengers, their luggage and cargo.

Left: Cleaning the windshield of a 747 before flight.

Fueling at contemporary airports can be accomplished by trucks accessing underground fuel depots. Fuel measurement is an important exercise. It affects the weight of the aircraft and its range. Fuel usage is one of the issues of nearly constant monitoring in the cockpit. World fuel prices have spiked twice in the last 20 years. In real terms fuel used in U.S. international service more than doubled in price in 1973. It rose more than 40 percent in the late 1970s. In each period of high fuel prices, world air travel dipped as did the world gross national product.

Left: The cockpit crew is busy doing last-minute checks on equipment and talking to the control tower, awaiting clearances to the next step in the process.

Opposite: Taxiing to Runway 13R/31L at JFK during rush hour.

Below: 747 landing gear and No. 2 and 3 engines. The amount of air taken in by each engine of a 747 may approach a ton per second. A gallon (3.78 liters) of fuel is consumed in under two seconds..

John F. Kennedy International Airport with the Pan American terminal and Runway 13R/31L from the air. The terminal now is owned by Delta Air Lines.

Above: The flight crew is prepped for flight by dispatchers. Much of the process has been computerized. A system has developed out of trial and error. It is focused on the weight of the aircraft, the fuel on board, the winds to be encountered. The cockpit crew checks the basic systems of the aircraft before pushback. A checklist is called out. Monitoring systems are available to warn the crew of the flight condition of the aircraft. At many airports, a tug is required to tow the aircraft to a point where the taxi may begin.

Left: The view from the cockpit of Runway 22L at JFK. U.S. pilots flying long-range transports are paid $150,000 a year and more. Duty hours are carefully controlled. Relief crews are required for long-range flights.

Takeoff on Runway 4L at JFK, viewed from cabin, when a fully loaded aircraft taxis for takeoff, one can tell that an aircraft was made to fly. The bulky machinery bumps along a taxiway, clumsily and slowly turning.

Once the clearance is given, the takeoff is a fuel-guzzling display of power. Takeoff speeds are the fastest that most people will ever achieve on the ground.

Engine power pulls the aircraft along to speeds at which the wing begins to work, lifting the machinery into the air. In the cockpit, the ride is more like floating at this point, up, up and up as the ground falls away.

After takeoff and climb out, the aircraft reaches a cruise altitude. These tracks are assigned by air traffic control and can be changed with approval. In the cockpit, the mandate is to see and avoid. Communications with air traffic control are regular and routine. Their job at ATC is maintaining separation of aircraft.

Flight at 33,000 feet on a 747. Most subsonic jets fly between 30,000 and 39,000 feet. Supersonic Concorde cruising at Mach 2 and between 52,000 and 59,000 feet allows one to see plainly the curvature of the earth from the window on a clear day

At cruise the average speed is 0.8 Mach, which is 80 percent of the speed of sound. Only the Anglo-French Concorde is capable of supersonic flight. Most aircraft on long-haul flights will be 7 to 8 miles above the earth. The Concorde flies higher, more than 10 miles above sea level.

L-1011 landing at New York's La Guardia. Supersonic Concorde lands at 187 mph (300 kph), slightly faster than the subsonic jets such as the 747, DC-10 and L-1011, which land at about 175 mph (285 kph).

Most of today's commercial airliners can land in bad weather, but there are limitations at many airports. In most cases, the pilots must have the ability to see the runway at an altitude at which they can decide to land or to go around. There are weather minimums in effect, below which most aircraft fly to alternate fields.

On approach there is one sound that all pilots want to hear. That is the "thrump-thrump" of the landing gear or undercarriage locking into position. Then the powered glide to the ground follows with a greater sense of security.

In today's aviation environment, aircraft begin their approaches to airports as far as 60 miles out. Coming from many directions, the aircraft are lined up by air traffic control. Their speed is ordered and their descent carefully watched.

Aircraft are the most maintained pieces of machinery in our world. Care of aircraft is as routine as the flight itself. The aircraft is monitored constantly as it operates. Details of malfunctions are kept in a log book. Entries are defined as essential and non-essential, but in each case they are action items at one point.

MAINTENANCE

DC-9-32 on a flight test after a D-check. Maintenance crews are available for line work. Many airlines refer to routine work as an A-check. There are more detailed checks referred to as B- and C-checks. A D-check is a complete overhaul, which is done every four years.

Hangars are large cavernous buildings, strangely clean for all the oil, kerosene and metalwork. Mechanics in overalls climb around the aircraft – looking, checking, fixing. Airplanes must be the most inspected and maintained vehicles in our world. The endless process starts each day with the first officer's pre-flight walkaround. Basic tools of flight are eyed. The wing flaps, ailerons, rudder, powerplants. And there are a variety of other routine checks, from A-level to D-level, from daily to every four to six years. Mechanics look at critical structures by removing access panels, even wing fairings. And they look deepest in a D-level heavy maintenance check, when the aircraft is inspected down to the bare metal for an assessment of its fundamental strength. Airlines write their own maintenance manuals. Aviation authorities approve of them. The chief targets of maintenance are metal fatigue and corrosion.

Computer technology has made aircraft surveillance more precise. Maintenance chiefs at a touch of the finger can pull onto a screen a list of airplanes in the fleet. They know when aircraft are due for a visit, when the components and systems should be checked and replaced. Directions for these tasks come from manufacturers and government aviation authorities. Signs of wear and corrosion prompt mechanics to repair or replacement. Fuselage skin can be repaired with covering doubles or skin patches, that overlap a damaged area. Whole fuselage panels are discarded and new skin applied, riveted, bonded and sealed. Old structural members are replaced. Metal is subjected to many kinds of tests for cracks, X-rays, dyes, electrical eddy current inspections. A heavy check every four years is so thorough that an aircraft can emerge after three weeks almost new.

Aircraft maintenance begins with a walk-around inspection and ends with maybe an eddy current inspection of a rivet. The key is to find the first crack.
– Ray Valeika former vice president, maintenance Pan American World Airways, currently head of maintenance at Continental Airlines.

In the heavy checks, all aircraft moveable parts are replaced. Thousands of man hours are consumed. Mechanics are given assignments based on an airline's maintenance manual, written to address issues that may arise in the route system flown by that airline, and the kind of operation it is – short-haul flying poses different problems from long-haul operations. Problems found by all concerned – the cockpit crew, the inspector, the mechanic – all are factored in a maintenance program. Airlines can spend up to $1 million on a refurbishment every four to five years, but the aircraft reemerges as a money-making flying machine. The debate over maintenance is rooted in economics. Older aircraft require more maintenance. Each airline must make the determination whether the investment in maintenance is worth it or whether the most economical course is to buy replacement aircraft.

Left: Most maintenance programs today are fed into computers. Heads of maintenance can tell at a glance which of their aircraft are due for checks and how long they will be out of service.

Below: Manufacturers produce service bulletins on aircraft and their systems. These bulletins recommend inspections and repair work and are based on information provided by the airline operators. Aviation authorities mandate some of these actions in airworthiness directives, which become a matter of regulation.

Above: Maintenance work on the carbon disk brakes of a Fokker 100. Disk brakes were used on aircraft long before cars.

Below: Repainting a 747 aircraft adds 1,000 lb. (450 kg.) to the weight. A smaller DC-9 such as this one gains about 250 lb. (114 kg.).

Right: Pratt & Whitney JT9D turbofans power the 747-100, -200 and -300 series. The JT9D was the first commercial high-bypass engine. The bypass ratio of air is five to six times that flown through the core engine

Top right: Inside the engine cowling is a noise attenuating lining

Far right: Rear engine mounting in the DC-10 is unusual. The engine is integrated with the tail fin, held there like a large cigar.

Right: A Pratt & Whitney JTD9 engine exhaust nozzle.

Right: L-1011-1 landing with Rolls-Royce RB211-22B turbofans. The more powerful RB211-524B engines were used on the more advanced versions of the Lockheed aircraft and on the 747-200 and -300.

AVIATION SAFETY

The pullout of the government from control of airline fares and routes – the trend referred to as airline deregulation – is not supposed to erode the level of air safety. Though it is difficult to pinpoint any adverse effects of deregulation on safety in the United States, where deregulation began, some definite flaws in the system have been isolated and much has changed for the better.

Everyone knows what air safety is – but do they? It may be defined as aviation without absolute accidents. Such a goal is an impossible one, as is absolute perfection in any area of human endeavor. Yet aviation with the fewest accidents is the goal of any air safety program. In terms of statistics, safety has tolerated deregulation. In the more than 10 years of deregulation in the United States, commercial aviation has had the fewest accidents of any period. For three years there were no fatal accidents at all involving large commercial airlines, the industry's best record.

Safety is a two-sided activity in aviation circles. First, it is a system of surveillance to find problems in aircraft, coupled with maintenance and repair to solve these problems. Secondly, it involves the adherence by pilots and others in the field to a set of rules that are codified after decades of hard and sometimes bitter experience. A healthy attitude is prevalent in aviation, based on learning from past lessons and honestly acknowledging deficiencies. Investigations of accidents and incidents seek a probable cause, not personal blame. These practices have resulted in a safe environment and the avoidance of countless adverse developments.

Aviation in itself is not inherently dangerous. But to an even greater degree than the sea it is terribly unforgiving of any carelessness, incapacity, or neglect.
– An advisory dating to World War II

Safety is best defined in terms of a person's exposure to risk. Mandatory steps to minimize risk are part and parcel of rules for an airline flight. The rules begin with the command structure. The pilot in charge is known as the P.I.C., the pilot in command. He must sign off on the fundamentals of flight: the proper amount of fuel, the route to be taken, and the usability and condition of the aircraft equipment.

Careful adherence to the rules is a basic requirement for safe operation, according to safety experts. A positive attitude toward safety begins with the top officials of the airlines.

To the safety experts, accident statistics are interesting guideposts, but they are not the final arbiter of an effective safety program or a guarantee of safe conditions. Accidents can happen to anyone. It is ironic that in this period of comparative safety in aviation, there are periods of acute public anxiety over an alleged lack of safety. This anxiety worsens after an accident and frequently sensational media reports.

The wind shear accidents of the 1980s captured the public attention as have few other accidents. Perhaps the apparent weather-related randomness of wind shear, yet the eerie regularity of a wind shear-related accident every three years, spurred the industry, the National Transportation Safety Board and the Federal Aviation Administration to action. Recognizing the hazards of wind shear, the Federal Aviation Administration developed a number of programs to better understand the phenomenon and to prepare pilots for handling wind shear encounters. By definition wind shear is a quick and dramatic change of wind direction, which can and frequently does have profound effects on airplane lift, particularly during takeoff or or landing.

Following the July, 1982, crash of a Pan American World Airways Boeing 727 at Kenner, Louisiana, just after take-off from New Orleans International Airport, the FAA placed the strongest emphasis on the problem of wind shear. A starting point for the FAA was basic research into wind shear conducted by University of Chicago professor Dr. Theodore Fujita. Additional research to understand wind shear and associated weather phenomena was undertaken by Dr. John McCarthy of the National Center for Atmospheric Research, in Boulder, Colorado. The effort begun in the post-1982 period resulted in a number of solutions that are now a part of the aviation safety system:
• A quick fix was a training aid for pilots that assisted in recognition of wind shear symptoms, with the basic advice to avoid it if at all possible. In the event that a wind shear encounter was about to occur, pilots are given a prescription for aircraft handling to ride it out.

- Improvements to the low-level wind shear alert system are now in operation at United States airports and others. This is the addition of wind shear-detecting sensors installed in airport perimeters. Detection of wind shifts are reported to the control tower. A system is being worked out so that pilots in the future will have wind shear alerts displayed for them on cockpit monitors.
- Doppler weather radar units have been installed at airports.
- Airborne wind shear detection devices based on the operating experience of Piedmont and United airlines also have been installed. The devices sense the effects of wind shear on the aircraft and provide the best means for escape. In the test phase, a Piedmont Boeing 737-200 outfitted with a Sperry, later Honeywel, airborne wind shear detection system escaped a shear encounter on approach to Chicago's O'Hare International Airport in November, 1985.

The captain of the Piedmont Boeing 737, alerted to the possibility of wind shear, noted the illumination of an amber warning light in the Sperry performance management system (PMS). Altitude was 1,000 feet and airspeed had increased to 170 knots. As a precaution the captain kept his twin engines spooled up high, and airspeed increased further. Extreme turbulence was encountered. The amber caution light shifted to red as the Boeing 737 descended to 600 feet and airspeed began to drop. The aircraft recorded a descent rate of 1,500 feet per minute. and was experiencing a 30-knot loss of airspeed. The pilot executed a go-around at maximum thrust and exited the wind shear.
– Report of a wind shear encounter in Aviation Week & Space Technology, September 22, 1986

Smugness about the good safety record since deregulation must be guarded against in order to maintain and foster a continuing awareness of the possibilities for critical errors. The necessity for accident prevention through the monitoring of incidents and modification of procedures as appropriate remains a cornerstone of safe operation.
– John H. Enders, President, Flight Safety Foundation, Arlington, Virginia

Neither the U.S. Civil Aeronautics Board nor its successor, the Transportation Department, have found any evidence of a deterioration in safety since the 1978 airline deregulation act. Annual reports to Congress by the agencies measured the impact of deregulation on safety by counting accidents, incidents and violations of regulations. In the decade since deregulation started, the accident record has remained basically the same or has improved in the various categories of airlines: commuter carriers, non-scheduled carriers and scheduled airlines.

Data maintained by the Flight Safety Foundation, Arlington, Virginia and the International Air Transport Association indicate that annual jet transport losses on a world basis have maintained the extraordinarily low level of one or two per million takeoff and landing cycles flown through much of the 1970s and 1980s.

An analysis of financial influences on airline safety has mixed implications for safety in a deregulated environment. Nancy L. Rose, assistant professor of applied economics at the Sloan School of Management, Massachusetts Institute of Technology, analyzed data on 31 airlines from 1955 through 1983 and found the aggregate safety performance of those airlines, as measured by accidents, "is superb and shows no sign of deterioration since deregulation." The record of three years under deregulation with no passenger fatalities was not equalled in any year prior to deregulation, she has pointed out. The study found evidence, however, that financial difficulties may be correlated with accident rates at individual airlines.

The U.S. government, alert to the economic pressures of deregulation and the heavy financial losses at some airlines, instituted a national inspection system of airlines of all categories in the middle 1980s. This new emphasis, involving formation of special investigative teams from FAA offices around the nation, resulted in millions of dollars in fines being levied against U.S. airlines for violations of FAA rules, including operating unairworthy aircraft, failing to carry out required maintenance and failing to keep proper records of the maintenance. In some cases, the discrepancies found at airlines resulted in the withdrawal of operating certificates. In most cases, the violations were not examples of gross neglect. The airlines properly cried foul to some of the charges. The FAA has been inconsistent in applying rules and regulations through its nine regions in the United States. The airlines, which grew to fully national operations in this period, found it difficult to operate under their redundant, inconsistent application of FAA rules.

The safety record of the world's airlines is an extraordinary one. Data made available by the Flight Safety Foundation shows that both fatal accidents and operational jet hull losses have been on a steadily declining course for decades. This is in part due to the redundancies in critical systems in aircraft that back up failure of primary systems. Fail-safe manufacturing design makes its own fundamental contribution to aviation safety. This concept means that aircraft are built to withstand failures and to absorb the shifts of weight loads from a failed component or structure to another.

A prime example, though a horrific one, is the midair partial breakup of an Aloha Airlines Boeing 737 in Spring, 1988, in the skies over Hawaii. The event galvanized world aviation authorities, airlines and manufacturers into a wholesale re-evaluation of the engineering effects of fatigue and corrosion on aging transports.

The top half of the 737's fuselage tube tore off the aircraft as it operated at 24,000-ft. altitude. The rapid decompression of the cabin resulted in the loss of one flight attendant who was swept away in the sudden blast. The passengers, advised by the cockpit crew to keep their seat belts buckled, survived the harrowing descent to an airport. Cracks and corrosion were discovered in the accident airplane and in sister 737s that had long flown in the salt-air environment of the islands. They were among the most used aircraft, approaching 90,000 cycles – a cycle being a landing and take-off, pressurization of the cabin and depressurization – of the 737 world fleet. Boeing had issued service bulletins advising special inspections of these high-time aircraft.

The Aloha incident has refocused attention of aviation authorities on the maintenance programs and ways of testing, through non-destructive means, the sturdiness of aircraft. These evaluations will have a large impact on the use of older aircraft in the future. Maintenance experts properly affirm that each aircraft is different from any other, though they are of the same type. Their operating environments differ as well, further complicating the application of one rule for a single type. One certain outcome of the Aloha incident is even greater attention to detail in maintenance. As one experienced airlines maintenance officer said, "It . . . certainly will make us more cognizant of the implications behind every bump or nick in the fuselage."

There is the possibility of a return to the concept of an aircraft having a "safe" life. Many aviation authorities had followed that concept, which allowed a specific operating life for an aircraft before it headed for the scrap heap. More likely is an even more thorough system of inspections and testing, and an extra-wary eye on high-cycle aircraft.

Some differences also exist among experts on the level of safety in proposals to extend the operating range of twin-engine transports. Use of twin-engine aircraft has expanded rapidly since Air Canada and Trans World Airlines, using Boeing 767s, started up scheduled transatlantic crossings in 1983. The Airbus A310 has joined the 767 as an extended-range transport. The venerable Boeing 737 first qualified for the special operating condition, exclusive to the South Pacific region.

The latest proposals would ease up on the restrictions on operating a twin-engine aircraft. Specifically, the twin jets would be able to fly on routes that take them more than two hours flying time from a suitable airport, at one-engine speed. The eased restrictions permit twin-engine extended-range operations in most areas of the world.

Extended twin operations exceeded one million flight hours in 1986, and are piling up rapidly as airlines begin new services from heartland U.S. cities to Europe, avoiding traditional gateways. Special attention has been given to the strict rules for extended-range flights, such as special training for crews and requirements in equipment to maintain critical systems in case of an engine problem.

Traditional safety concerns have come to the fore. What of the effects of weather on aircraft operation and uses of airports? What are the human factors involved? What is the level of risk?

Statistics are impressively in favor of extending the range of twin-engine transports. Most major U.S. airlines are flying hundreds of extended-range flights each month without incident. The few incidents that have occurred, in which an aircraft had to divert to a suitable airport under one-engine power, have been investigated thoroughly. It appears that more extended-range flights are in our future.

Appendix
JET TRANSPORT CHARACTERISTICS

AEROSPATIALE/BRITISH AEROSPACE	**CONCORDE**
AIRBUS INDUSTRIE	A300
	A310
	A320
	A321
	A330
	A340
BOEING	707
	720
	727
	737
	747
	757
	767
BRITISH AEROSPACE	BAC 111
	HS121 TRIDENT
	VC10
	BAe 146
CONVAIR	CV880
	CV990
DASSAULT-BREGUET	MERCURE 100
de HAVILLAND	**DH106 COMET**
FOKKER	F28
	FOKKER 100
ILYUSHIN	IL-62
	IL-86
	IL-96
LOCKHEED	L-1011 TRISTAR
McDONNELL DOUGLAS	DC-8
	DC-9
	MD-80
	MD-87
	DC-10
	MD-11
SUD-EST	SE210 CARAVELLE
TUPOLEV	TU-104
	TU-124
	TU-134
	TU-144
	TU-154
	TU-204

Concorde in British Airways livery used from 1974 to 1985

AEROSPATIALE/BAe CONCORDE

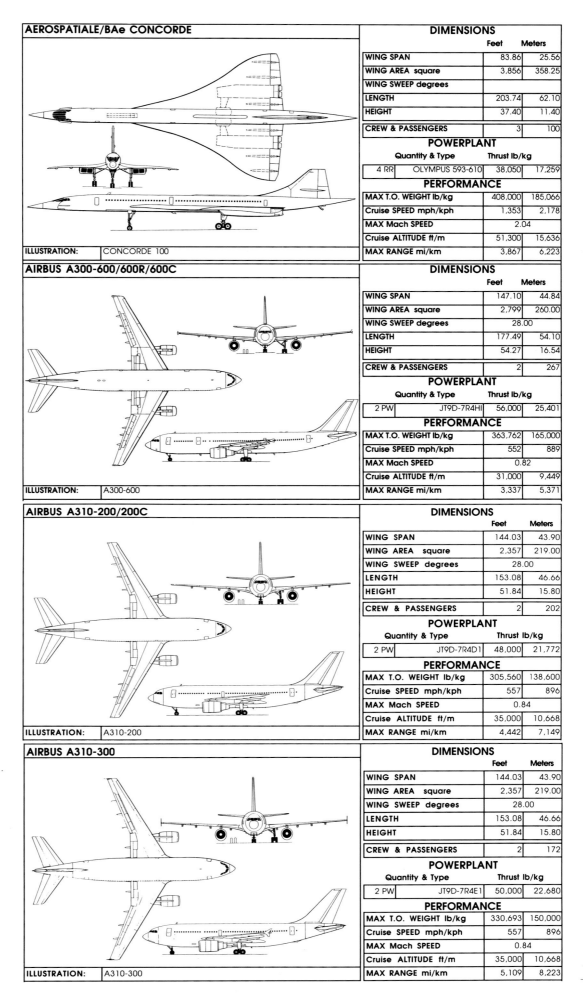

ILLUSTRATION: CONCORDE 100

DIMENSIONS	Feet	Meters
WING SPAN	83.86	25.56
WING AREA square	3,856	358.25
WING SWEEP degrees		
LENGTH	203.74	62.10
HEIGHT	37.40	11.40
CREW & PASSENGERS	3	100
POWERPLANT		
Quantity & Type	Thrust lb/kg	
4 RR OLYMPUS 593-610	38,050	17,259
PERFORMANCE		
MAX T.O. WEIGHT lb/kg	408,000	185,066
Cruise SPEED mph/kph	1,353	2,178
MAX Mach SPEED	2.04	
Cruise ALTITUDE ft/m	51,300	15,636
MAX RANGE mi/km	3,867	6,223

AIRBUS A300-600/600R/600C

ILLUSTRATION: A300-600

DIMENSIONS	Feet	Meters
WING SPAN	147.10	44.84
WING AREA square	2,799	260.00
WING SWEEP degrees	28.00	
LENGTH	177.49	54.10
HEIGHT	54.27	16.54
CREW & PASSENGERS	2	267
POWERPLANT		
Quantity & Type	Thrust lb/kg	
2 PW JT9D-7R4H1	56,000	25,401
PERFORMANCE		
MAX T.O. WEIGHT lb/kg	363,762	165,000
Cruise SPEED mph/kph	552	889
MAX Mach SPEED	0.82	
Cruise ALTITUDE ft/m	31,000	9,449
MAX RANGE mi/km	3,337	5,371

AIRBUS A310-200/200C

ILLUSTRATION: A310-200

DIMENSIONS	Feet	Meters
WING SPAN	144.03	43.90
WING AREA square	2,357	219.00
WING SWEEP degrees	28.00	
LENGTH	153.08	46.66
HEIGHT	51.84	15.80
CREW & PASSENGERS	2	202
POWERPLANT		
Quantity & Type	Thrust lb/kg	
2 PW JT9D-7R4D1	48,000	21,772
PERFORMANCE		
MAX T.O. WEIGHT lb/kg	305,560	138,600
Cruise SPEED mph/kph	557	896
MAX Mach SPEED	0.84	
Cruise ALTITUDE ft/m	35,000	10,668
MAX RANGE mi/km	4,442	7,149

AIRBUS A310-300

ILLUSTRATION: A310-300

DIMENSIONS	Feet	Meters
WING SPAN	144.03	43.90
WING AREA square	2,357	219.00
WING SWEEP degrees	28.00	
LENGTH	153.08	46.66
HEIGHT	51.84	15.80
CREW & PASSENGERS	2	172
POWERPLANT		
Quantity & Type	Thrust lb/kg	
2 PW JT9D-7R4E1	50,000	22,680
PERFORMANCE		
MAX T.O. WEIGHT lb/kg	330,693	150,000
Cruise SPEED mph/kph	557	896
MAX Mach SPEED	0.84	
Cruise ALTITUDE ft/m	35,000	10,668
MAX RANGE mi/km	5,109	8,223

AIRBUS A320-100/200

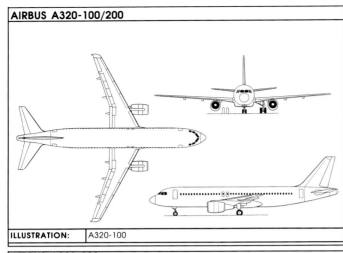

ILLUSTRATION: A320-100

DIMENSIONS	Feet	Meters
WING SPAN	111.25	33.91
WING AREA square	1,320	122.60
WING SWEEP degrees	25.00	
LENGTH	123.26	37.57
HEIGHT	38.58	11.76
CREW & PASSENGERS	2	150

POWERPLANT		
Quantity & Type	Thrust lb/kg	
2 CFMI CFM56-5A1	25,000	11,340

PERFORMANCE		
MAX T.O. WEIGHT lb/kg	149,914	68,000
Cruise SPEED mph/kph	522	841
MAX Mach SPEED	0.82	
Cruise ALTITUDE ft/m	37,000	11,278
MAX RANGE mi/km	2,140	3,445

AIRBUS A321-100

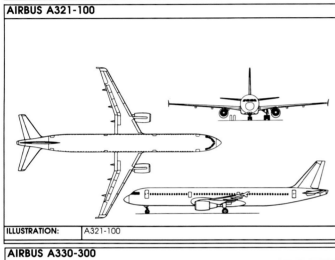

ILLUSTRATION: A321-100

DIMENSIONS	Feet	Meters
WING SPAN	111.88	34.10
WING AREA square	1,320	122.60
WING SWEEP degrees	25.00	
LENGTH	146.03	44.51
HEIGHT	38.75	11.81
CREW & PASSENGERS	2	180

POWERPLANT		
Quantity & Type	Thrust lb/kg	
2 CFMI CFM56-5B	30,000	13,608

PERFORMANCE		
MAX T.O. WEIGHT lb/kg	182,983	82,000
Cruise SPEED mph/kph	560	902
MAX Mach SPEED	0.82	
Cruise ALTITUDE ft/m	28,000	8,534
MAX RANGE mi/km	2,265	3,645

AIRBUS A330-300

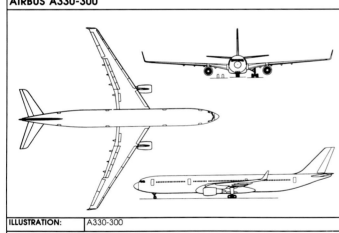

ILLUSTRATION: A330-300

DIMENSIONS	Feet	Meters
WING SPAN	197.83	60.30
WING AREA square	3,897	362.00
WING SWEEP degrees	30.00	
LENGTH	208.66	63.60
HEIGHT	54.79	16.70
CREW & PASSENGERS	2	335

POWERPLANT		
Quantity & Type	Thrust lb/kg	
2 GE CF6-80C2E1A1	67,500	30,617

PERFORMANCE		
MAX T.O. WEIGHT lb/kg	467,379	212,000
Cruise SPEED mph/kph	575	926
MAX Mach SPEED	0.86	
Cruise ALTITUDE ft/m	33,000	10,058
MAX RANGE mi/km	3,999	6,436

AIRBUS A340-300/300 Combi

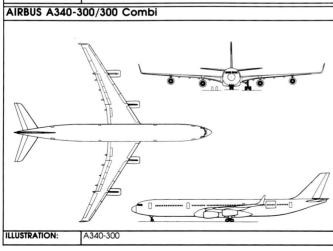

ILLUSTRATION: A340-300

DIMENSIONS	Feet	Meters
WING SPAN	197.83	60.30
WING AREA square	3,897	362.00
WING SWEEP degrees	30.00	
LENGTH	208.66	63.60
HEIGHT	54.79	16.70
CREW & PASSENGERS	2	295

POWERPLANT		
Quantity & Type	Thrust lb/kg	
4 CFMI CFM56-5C2	31,200	14,152

PERFORMANCE		
MAX T.O. WEIGHT lb/kg	588,871	253,500
Cruise SPEED mph/kph	575	926
MAX Mach SPEED	0.86	
Cruise ALTITUDE ft/m	33,000	10,058
MAX RANGE mi/km	6,741	10,849

BOEING 707-320B/320C

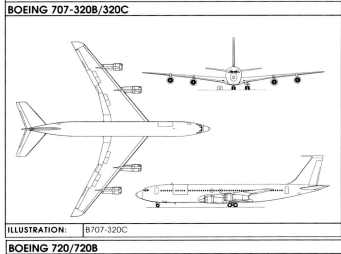

ILLUSTRATION: B707-320C

DIMENSIONS	Feet	Meters
WING SPAN	145.73	44.42
WING AREA square	2,892	268.70
WING SWEEP degrees	35.00	
LENGTH	152.92	46.61
HEIGHT	42.42	12.93
CREW & PASSENGERS	3	149

POWERPLANT		
Quantity & Type		Thrust lb/kg
4 PW	JT3D-3/3B	18,000 8,165

PERFORMANCE		
MAX T.O. WEIGHT lb/kg	333,592	151,315
Cruise SPEED mph/kph	627	1,009
MAX Mach SPEED	0.95	
Cruise ALTITUDE ft/m	39,000	11,887
MAX RANGE mi/km	4,948	7,964

BOEING 720/720B

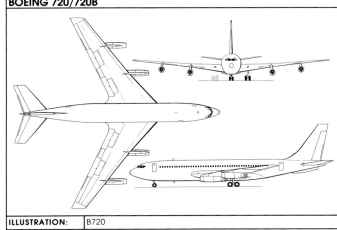

ILLUSTRATION: B720

DIMENSIONS	Feet	Meters
WING SPAN	130.84	39.88
WING AREA square	2,433	226.00
WING SWEEP degrees	35.00	
LENGTH	136.75	41.68
HEIGHT	41.57	12.67
CREW & PASSENGERS	3	124

POWERPLANT		
Quantity & Type		Thrust lb/kg
4 PW	JT3D-1	17,000 7,711

PERFORMANCE		
MAX T.O. WEIGHT lb/kg	234,001	106,141
Cruise SPEED mph/kph	604	972
MAX Mach SPEED	0.95	
Cruise ALTITUDE ft/m	25,000	7,620
MAX RANGE mi/km	4,730	7,612

BOEING 727-200/200F

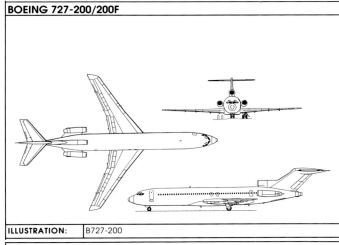

ILLUSTRATION: B727-200

DIMENSIONS	Feet	Meters
WING SPAN	108.01	32.92
WING AREA square	1,560	144.90
WING SWEEP degrees	32.00	
LENGTH	153.18	46.69
HEIGHT	33.99	10.36
CREW & PASSENGERS	3	134

POWERPLANT		
Quantity & Type		Thrust lb/kg
3 PW	JT8D-11	15,000 6,804

PERFORMANCE		
MAX T.O. WEIGHT lb/kg	174,826	79,300
Cruise SPEED mph/kph	598	963
MAX Mach SPEED	0.90	
Cruise ALTITUDE ft/m	24,700	7,529
MAX RANGE mi/km	2,463	3,963

BOEING 737-200/200C/QC

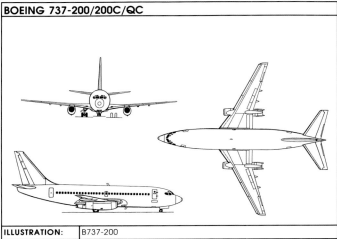

ILLUSTRATION: B737-200

DIMENSIONS	Feet	Meters
WING SPAN	92.85	28.30
WING AREA square	980	91.00
WING SWEEP degrees	25.00	
LENGTH	100.16	30.53
HEIGHT	37.01	11.28
CREW & PASSENGERS	2	102

POWERPLANT		
Quantity & Type		Thrust lb/kg
2 PW	JT8D-15A	15,500 7,031

PERFORMANCE		
MAX T.O. WEIGHT lb/kg	115,522	52,400
Cruise SPEED mph/kph	562	904
MAX Mach SPEED	0.84	
Cruise ALTITUDE ft/m	25,000	7,620
MAX RANGE mi/km	2,135	3,435

BOEING 737-400

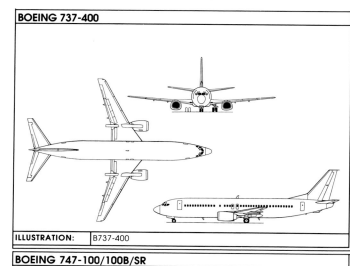

ILLUSTRATION: B737-400

DIMENSIONS		
	Feet	Meters
WING SPAN	94.82	28.90
WING AREA square	980	91.00
WING SWEEP degrees	25.00	
LENGTH	119.42	36.40
HEIGHT	36.42	11.10
CREW & PASSENGERS	2	146
POWERPLANT		
Quantity & Type	Thrust lb/kg	
2 CFMI CFM56-3B2	22,000	9,979
PERFORMANCE		
MAX T.O. WEIGHT lb/kg	138,494	62,820
Cruise SPEED mph/kph	566	911
MAX Mach SPEED	0.82	
Cruise ALTITUDE ft/m	26,000	7,925
MAX RANGE mi/km	2,244	3,611

BOEING 747-100/100B/SR

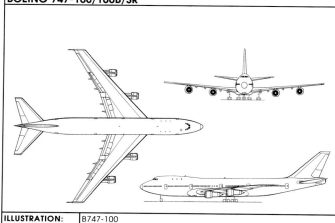

ILLUSTRATION: B747-100

DIMENSIONS		
	Feet	Meters
WING SPAN	195.67	59.64
WING AREA square	5,500	510.97
WING SWEEP degrees	37.50	
LENGTH	231.82	70.66
HEIGHT	63.42	19.33
CREW & PASSENGERS	3	385
POWERPLANT		
Quantity & Type	Thrust lb/kg	
4 PW JT9D-7A	46,960	21,301
PERFORMANCE		
MAX T.O. WEIGHT lb/kg	735,000	333,391
Cruise SPEED mph/kph	583	939
MAX Mach SPEED	0.92	
Cruise ALTITUDE ft/m	35,000	10,668
MAX RANGE mi/km	5,098	8,204

BOEING 747-200B/C/F/Combi

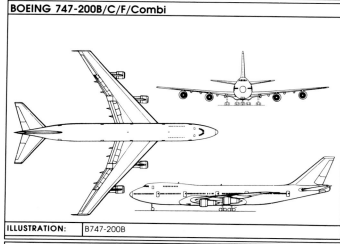

ILLUSTRATION: B747-200B

DIMENSIONS		
	Feet	Meters
WING SPAN	195.67	59.64
WING AREA square	5,500	510.97
WING SWEEP degrees	37.50	
LENGTH	231.82	70.66
HEIGHT	63.42	19.33
CREW & PASSENGERS	3	366
POWERPLANT		
Quantity & Type	Thrust lb/kg	
4 GE CF6-50E2	52,500	23,814
PERFORMANCE		
MAX T.O. WEIGHT lb/kg	800,000	362,875
Cruise SPEED mph/kph	583	939
MAX Mach SPEED	0.92	
Cruise ALTITUDE ft/m	35,000	10,668
MAX RANGE mi/km	6,836	11,001

BOEING 747-300/300SR/Combi

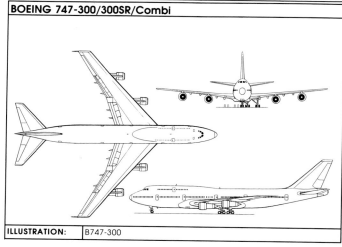

ILLUSTRATION: B747-300

DIMENSIONS		
	Feet	Meters
WING SPAN	195.67	59.64
WING AREA square	5,500	510.97
WING SWEEP degrees	37.50	
LENGTH	231.82	70.66
HEIGHT	63.42	19.33
CREW & PASSENGERS	3	400
POWERPLANT		
Quantity & Type	Thrust lb/kg	
4 PW JT9D-7R4G2	54,750	24,834
PERFORMANCE		
MAX T.O. WEIGHT lb/kg	820,000	371,946
Cruise SPEED mph/kph	557	896
MAX Mach SPEED	0.92	
Cruise ALTITUDE ft/m	35,000	10,668
MAX RANGE mi/km	6,525	10,501

BOEING 747-400/400 Combi

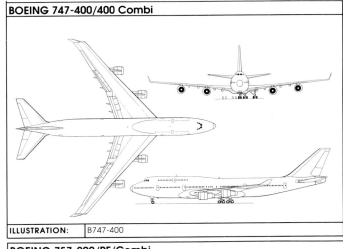

ILLUSTRATION: B747-400

DIMENSIONS	Feet	Meters
WING SPAN	210.99	64.31
WING AREA square	5,650	524.90
WING SWEEP degrees	37.50	
LENGTH	231.82	70.66
HEIGHT	63.42	19.33
CREW & PASSENGERS	2	412
POWERPLANT		
Quantity & Type	Thrust lb/kg	
4 PW PW4256	56,000	25,401
PERFORMANCE		
MAX T.O. WEIGHT lb/kg	870,000	394,626
Cruise SPEED mph/kph	583	939
MAX Mach SPEED	0.92	
Cruise ALTITUDE ft/m	35,000	10,668
MAX RANGE mi/km	7,998	12,871

BOEING 757-200/PF/Combi

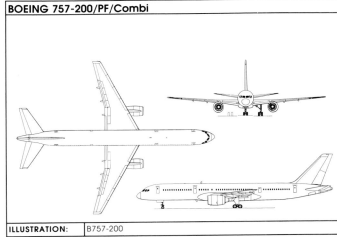

ILLUSTRATION: B757-200

DIMENSIONS	Feet	Meters
WING SPAN	124.84	38.05
WING AREA square	1,950	181.20
WING SWEEP degrees	25.00	
LENGTH	155.25	47.32
HEIGHT	44.49	13.56
CREW & PASSENGERS	2	186
POWERPLANT		
Quantity & Type	Thrust lb/kg	
2 RR RB211-535C	37,400	16,964
PERFORMANCE		
MAX T.O. WEIGHT lb/kg	240,000	108,863
Cruise SPEED mph/kph	528	850
MAX Mach SPEED	0.86	
Cruise ALTITUDE ft/m	39,000	11,887
MAX RANGE mi/km	3,659	5,889

BOEING 767-200/200ER

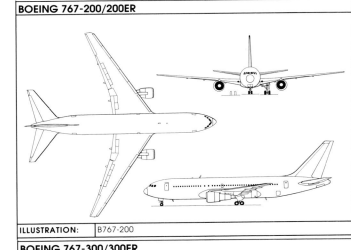

ILLUSTRATION: B767-200

DIMENSIONS	Feet	Meters
WING SPAN	156.07	47.57
WING AREA square	3,050	283.35
WING SWEEP degrees	31.50	
LENGTH	159.15	48.51
HEIGHT	52.00	15.85
CREW & PASSENGERS	2	216
POWERPLANT		
Quantity & Type	Thrust lb/kg	
2 PW JT9D-7R4D	48,000	21,772
PERFORMANCE		
MAX T.O. WEIGHT lb/kg	300,000	136,078
Cruise SPEED mph/kph	527	848
MAX Mach SPEED	0.86	
Cruise ALTITUDE ft/m	39,000	11,887
MAX RANGE mi/km	3,671	5,908

BOEING 767-300/300ER

ILLUSTRATION: B767-300

DIMENSIONS	Feet	Meters
WING SPAN	156.07	47.57
WING AREA square	3,050	283.35
WING SWEEP degrees	31.50	
LENGTH	180.12	54.90
HEIGHT	52.00	15.85
CREW & PASSENGERS	2	261
POWERPLANT		
Quantity & Type	Thrust lb/kg	
2 PW JT9D-7R4D	48,000	21,772
PERFORMANCE		
MAX T.O. WEIGHT lb/kg	345,000	156,490
Cruise SPEED mph/kph	528	850
MAX Mach SPEED	0.86	
Cruise ALTITUDE ft/m	39,000	11,887
MAX RANGE mi/km	3,717	5,982

BAe (BAC)111-500

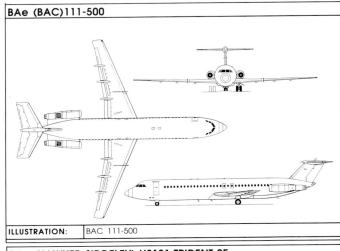

ILLUSTRATION: BAC 111-500

DIMENSIONS	Feet	Meters
WING SPAN	93.50	28.50
WING AREA square	1,031	95.78
WING SWEEP degrees	20.00	
LENGTH	106.99	32.61
HEIGHT	24.51	7.47
CREW & PASSENGERS	2	89

POWERPLANT		
Quantity & Type	Thrust lb/kg	
2 RR RB163 SPEY Mk512	12,000	5,443

PERFORMANCE		
MAX T.O. WEIGHT lb/kg	96,011	43,550
Cruise SPEED mph/kph	526	846
MAX Mach SPEED	0.78	
Cruise ALTITUDE ft/m	30,000	9,144
MAX RANGE mi/km	1,703	2,741

BAe (HAWKER-SIDDELEY) HS121 TRIDENT 2E

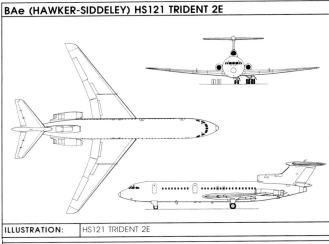

ILLUSTRATION: HS121 TRIDENT 2E

DIMENSIONS	Feet	Meters
WING SPAN	98.00	29.87
WING AREA square	1,456	135.26
WING SWEEP degrees	35.00	
LENGTH	114.76	34.98
HEIGHT	27.00	8.23
CREW & PASSENGERS	3	115

POWERPLANT		
Quantity & Type	Thrust lb/kg	
3 RR RB163 SPEY Mk512	11,930	5,411

PERFORMANCE		
MAX T.O. WEIGHT lb/kg	143,500	65,091
Cruise SPEED mph/kph	605	974
MAX Mach SPEED	0.95	
Cruise ALTITUDE ft/m	27,000	8,230
MAX RANGE mi/km	3,631	5,843

BAe (VICKERS) VC10-1151 SUPER

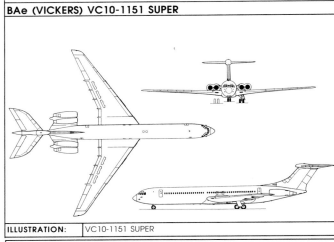

ILLUSTRATION: VC10-1151 SUPER

DIMENSIONS	Feet	Meters
WING SPAN	158.66	48.36
WING AREA square	2,932	272.40
WING SWEEP degrees	37.00	
LENGTH	158.66	48.36
HEIGHT	39.50	12.04
CREW & PASSENGERS	3	174

POWERPLANT		
Quantity & Type	Thrust lb/kg	
4 RR CONWAY 43 Mk550	21,800	9,888

PERFORMANCE		
MAX T.O. WEIGHT lb/kg	334,992	151,950
Cruise SPEED mph/kph	581	935
MAX Mach SPEED	0.86	
Cruise ALTITUDE ft/m	31,000	9,449
MAX RANGE mi/km	4,488	7,223

BAe 146-200/200QT/QC

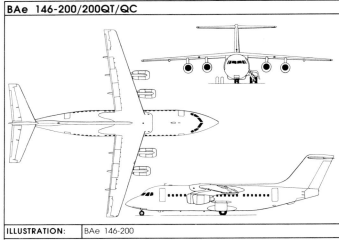

ILLUSTRATION: BAe 146-200

DIMENSIONS	Feet	Meters
WING SPAN	86.42	26.34
WING AREA square	832	77.30
WING SWEEP degrees	15.00	
LENGTH	93.83	28.60
HEIGHT	28.25	8.61
CREW & PASSENGERS	2	74

POWERPLANT		
Quantity & Type	Thrust lb/kg	
4AVCO ALF 502R-3A/5	6,968	3,161

PERFORMANCE		
MAX T.O. WEIGHT lb/kg	93,000	42,184
Cruise SPEED mph/kph	487	783
MAX Mach SPEED	0.70	
Cruise ALTITUDE ft/m	24,000	7,315
MAX RANGE mi/km	1,855	2,985

BAe 146-300/300QT/QC

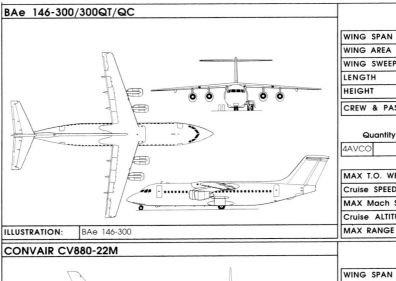

ILLUSTRATION: BAe 146-300

DIMENSIONS	Feet	Meters
WING SPAN	86.42	26.34
WING AREA square	829	77.00
WING SWEEP degrees	15.00	
LENGTH	101.67	30.99
HEIGHT	28.25	8.61
CREW & PASSENGERS	2	103
POWERPLANT		
Quantity & Type	Thrust lb/kg	
4 AVCO ALF 502R-3A/5	6,968	3,161
PERFORMANCE		
MAX T.O. WEIGHT lb/kg	95,001	43,092
Cruise SPEED mph/kph	428	689
MAX Mach SPEED	0.70	
Cruise ALTITUDE ft/m	31,000	9,449
MAX RANGE mi/km	1,587	2,554

CONVAIR CV880-22M

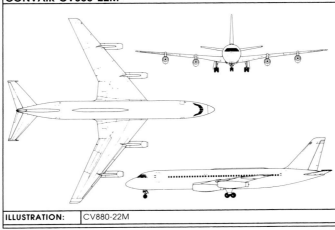

ILLUSTRATION: CV880-22M

DIMENSIONS	Feet	Meters
WING SPAN	120.01	36.58
WING AREA square	2,000	185.80
WING SWEEP degrees	35.00	
LENGTH	129.33	39.42
HEIGHT	36.09	11.00
CREW & PASSENGERS	4	84
POWERPLANT		
Quantity & Type	Thrust lb/kg	
4 GE CJ805-3B tj	11,650	5,284
PERFORMANCE		
MAX T.O. WEIGHT lb/kg	193,499	87,770
Cruise SPEED mph/kph	616	991
MAX Mach SPEED	0.82	
Cruise ALTITUDE ft/m	22,500	6,858
MAX RANGE mi/km	2,877	4,630

CONVAIR CV990-30A

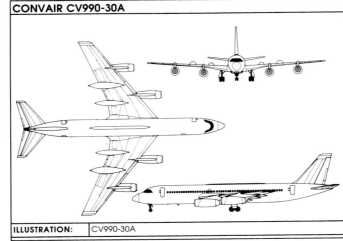

ILLUSTRATION: CV990-30A

DIMENSIONS	Feet	Meters
WING SPAN	120.01	36.58
WING AREA square	2,250	209.00
WING SWEEP degrees	35.00	
LENGTH	139.21	42.43
HEIGHT	39.50	12.04
CREW & PASSENGERS	3	97
POWERPLANT		
Quantity & Type	Thrust lb/kg	
4 GE CJ805-23B tj	16,050	7,280
PERFORMANCE		
MAX T.O. WEIGHT lb/kg	253,002	114,760
Cruise SPEED mph/kph	615	990
MAX Mach SPEED	0.87	
Cruise ALTITUDE ft/m	20,000	6,096
MAX RANGE mi/km	3,063	4,930

DASSAULT-BREGUET MERCURE 100

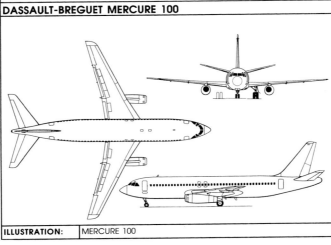

ILLUSTRATION: MERCURE 100

DIMENSIONS	Feet	Meters
WING SPAN	100.23	30.55
WING AREA square	1,249	116.00
WING SWEEP degrees	25.00	
LENGTH	114.30	34.84
HEIGHT	37.27	11.36
CREW & PASSENGERS	2	120
POWERPLANT		
Quantity & Type	Thrust lb/kg	
2 PW JT8D-15	15,500	7,031
PERFORMANCE		
MAX T.O. WEIGHT lb/kg	124,561	56,500
Cruise SPEED mph/kph	575	926
MAX Mach SPEED	0.85	
Cruise ALTITUDE ft/m	20,000	6,096
MAX RANGE mi/km	1,283	2,065

de HAVILLAND DH106 COMET 4C

ILLUSTRATION: DH106 COMET 4C

DIMENSIONS	Feet	Meters
WING SPAN	114.83	35.00
WING AREA square	2,121	197.05
WING SWEEP degrees	20.00	
LENGTH	118.00	35.97
HEIGHT	29.43	8.97
CREW & PASSENGERS	3	60-72
POWERPLANT		
Quantity & Type	Thrust lb/kg	
4 RR AVON 525B	10,500	4,763
PERFORMANCE		
MAX T.O. WEIGHT lb/kg	162,000	73,482
Cruise SPEED mph/kph	503	809
MAX Mach SPEED	0.73	
Cruise ALTITUDE ft/m	39,000	11,887
MAX RANGE mi/km	2,681	4,315

FOKKER F28-4000

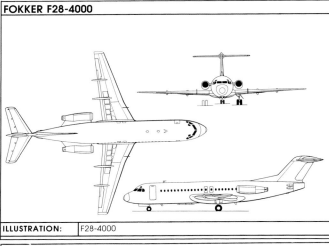

ILLUSTRATION: F28-4000

DIMENSIONS	Feet	Meters
WING SPAN	82.25	25.07
WING AREA square	821	76.30
WING SWEEP degrees	16.00	
LENGTH	97.11	29.60
HEIGHT	27.79	8.47
CREW & PASSENGERS	2	75
POWERPLANT		
Quantity & Type	Thrust lb/kg	
2 RR RB183-Mk555-15P	9,900	4,491
PERFORMANCE		
MAX T.O. WEIGHT lb/kg	69,496	31,523
Cruise SPEED mph/kph	502	807
MAX Mach SPEED	0.75	
Cruise ALTITUDE ft/m	33,000	10,058
MAX RANGE mi/km	1,083	1,743

FOKKER 100

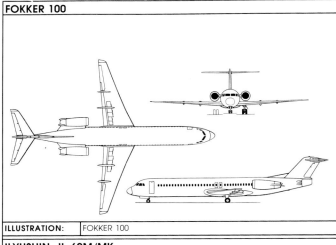

ILLUSTRATION: FOKKER 100

DIMENSIONS	Feet	Meters
WING SPAN	92.13	28.08
WING AREA square	1,015	94.30
WING SWEEP degrees	17.45	
LENGTH	116.57	35.53
HEIGHT	27.89	8.50
CREW & PASSENGERS	2	85
POWERPLANT		
Quantity & Type	Thrust lb/kg	
2 RR TAY 620-15	13,850	6,282
PERFORMANCE		
MAX T.O. WEIGHT lb/kg	95,000	43,092
Cruise SPEED mph/kph	525	845
MAX Mach SPEED	0.77	
Cruise ALTITUDE ft/m	27,000	8,230
MAX RANGE mi/km	1,343	2,161

ILYUSHIN IL-62M/MK

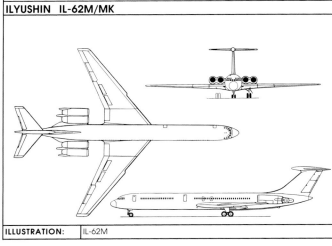

ILLUSTRATION: IL-62M

DIMENSIONS	Feet	Meters
WING SPAN	141.73	43.20
WING AREA square	3,009	279.55
WING SWEEP degrees	32.30	
LENGTH	174.28	53.12
HEIGHT	40.52	12.35
CREW & PASSENGERS	5	140
POWERPLANT		
Quantity & Type	Thrust lb/kg	
4 SOLOVIEV D-30KU	24,250	11,000
PERFORMANCE		
MAX T.O. WEIGHT lb/kg	368,172	167,000
Cruise SPEED mph/kph	571	919
MAX Mach SPEED	0.83	
Cruise ALTITUDE ft/m	26,200	7,986
MAX RANGE mi/km	4,845	7,797

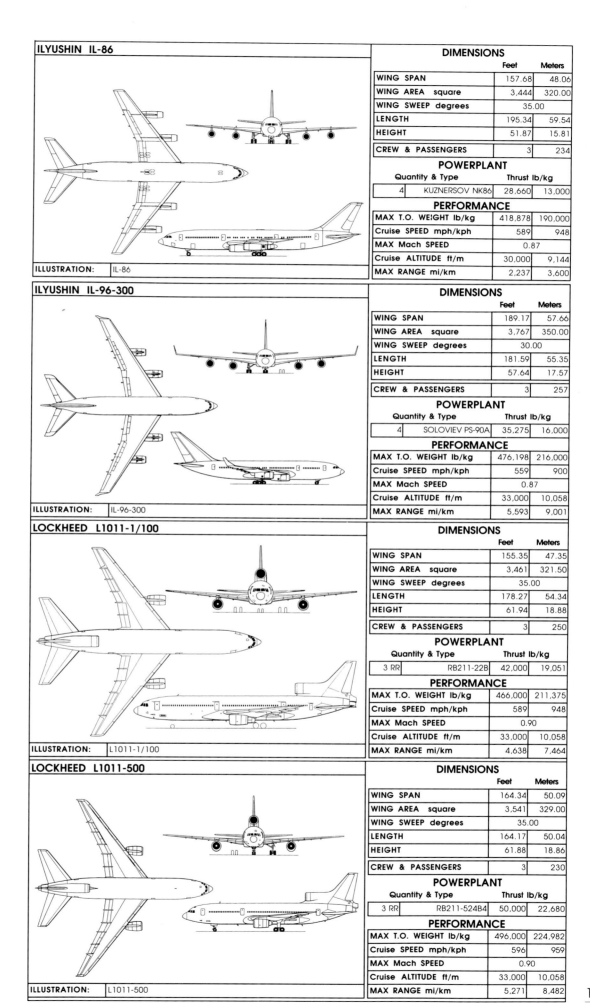

ILYUSHIN IL-86

ILLUSTRATION: IL-86

DIMENSIONS	Feet	Meters
WING SPAN	157.68	48.06
WING AREA square	3,444	320.00
WING SWEEP degrees	35.00	
LENGTH	195.34	59.54
HEIGHT	51.87	15.81
CREW & PASSENGERS	3	234

POWERPLANT		
Quantity & Type	Thrust lb/kg	
4 KUZNERSOV NK86	28,660	13,000

PERFORMANCE		
MAX T.O. WEIGHT lb/kg	418,878	190,000
Cruise SPEED mph/kph	589	948
MAX Mach SPEED	0.87	
Cruise ALTITUDE ft/m	30,000	9,144
MAX RANGE mi/km	2,237	3,600

ILYUSHIN IL-96-300

ILLUSTRATION: IL-96-300

DIMENSIONS	Feet	Meters
WING SPAN	189.17	57.66
WING AREA square	3,767	350.00
WING SWEEP degrees	30.00	
LENGTH	181.59	55.35
HEIGHT	57.64	17.57
CREW & PASSENGERS	3	257

POWERPLANT		
Quantity & Type	Thrust lb/kg	
4 SOLOVIEV PS-90A	35,275	16,000

PERFORMANCE		
MAX T.O. WEIGHT lb/kg	476,198	216,000
Cruise SPEED mph/kph	559	900
MAX Mach SPEED	0.87	
Cruise ALTITUDE ft/m	33,000	10,058
MAX RANGE mi/km	5,593	9,001

LOCKHEED L1011-1/100

ILLUSTRATION: L1011-1/100

DIMENSIONS	Feet	Meters
WING SPAN	155.35	47.35
WING AREA square	3,461	321.50
WING SWEEP degrees	35.00	
LENGTH	178.27	54.34
HEIGHT	61.94	18.88
CREW & PASSENGERS	3	250

POWERPLANT		
Quantity & Type	Thrust lb/kg	
3 RR RB211-22B	42,000	19,051

PERFORMANCE		
MAX T.O. WEIGHT lb/kg	466,000	211,375
Cruise SPEED mph/kph	589	948
MAX Mach SPEED	0.90	
Cruise ALTITUDE ft/m	33,000	10,058
MAX RANGE mi/km	4,638	7,464

LOCKHEED L1011-500

ILLUSTRATION: L1011-500

DIMENSIONS	Feet	Meters
WING SPAN	164.34	50.09
WING AREA square	3,541	329.00
WING SWEEP degrees	35.00	
LENGTH	164.17	50.04
HEIGHT	61.88	18.86
CREW & PASSENGERS	3	230

POWERPLANT		
Quantity & Type	Thrust lb/kg	
3 RR RB211-524B4	50,000	22,680

PERFORMANCE		
MAX T.O. WEIGHT lb/kg	496,000	224,982
Cruise SPEED mph/kph	596	959
MAX Mach SPEED	0.90	
Cruise ALTITUDE ft/m	33,000	10,058
MAX RANGE mi/km	5,271	8,482

DOUGLAS DC8-30

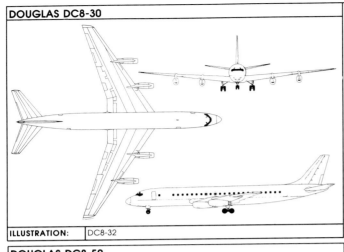

ILLUSTRATION: DC8-32

DIMENSIONS	Feet	Meters
WING SPAN	142.42	43.41
WING AREA square	2,771	257.44
WING SWEEP degrees	30.60	
LENGTH	150.49	45.87
HEIGHT	43.34	13.21
CREW & PASSENGERS	4	132

POWERPLANT		
Quantity & Type		Thrust lb/kg
4 PW	JT4A-11 tj	17,500	7,938

PERFORMANCE		
MAX T.O. WEIGHT lb/kg	310,000	140,614
Cruise SPEED mph/kph	589	948
MAX Mach SPEED	0.88	
Cruise ALTITUDE ft/m	31,000	9,449
MAX RANGE mi/km	4,638	7,464

DOUGLAS DC8-50

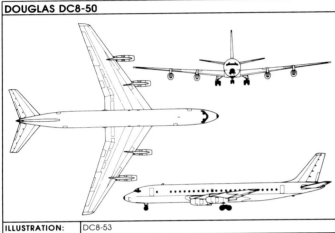

ILLUSTRATION: DC8-53

DIMENSIONS	Feet	Meters
WING SPAN	142.42	43.41
WING AREA square	2,883	267.84
WING SWEEP degrees	30.60	
LENGTH	150.49	45.87
HEIGHT	43.34	13.21
CREW & PASSENGERS	4	150

POWERPLANT		
Quantity & Type		Thrust lb/kg
4 PW	JT3D-3	18,000	8,165

PERFORMANCE		
MAX T.O. WEIGHT lb/kg	300,000	136,078
Cruise SPEED mph/kph	593	954
MAX Mach SPEED	0.88	
Cruise ALTITUDE ft/m	31,000	9,449
MAX RANGE mi/km	4,546	7,315

DOUGLAS DC8-61

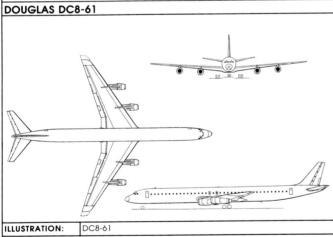

ILLUSTRATION: DC8-61

DIMENSIONS	Feet	Meters
WING SPAN	142.42	43.41
WING AREA square	2,883	267.84
WING SWEEP degrees	30.60	
LENGTH	187.34	57.10
HEIGHT	43.34	13.21
CREW & PASSENGERS	3	220

POWERPLANT		
Quantity & Type		Thrust lb/kg
4 PW	JT3D-3B	18,000	8,165

PERFORMANCE		
MAX T.O. WEIGHT lb/kg	325,000	147,418
Cruise SPEED mph/kph	581	935
MAX Mach SPEED	0.88	
Cruise ALTITUDE ft/m	31,000	9,449
MAX RANGE mi/km	3,372	5,426

DOUGLAS DC8-62

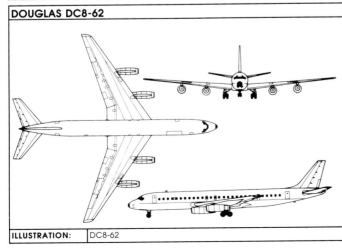

ILLUSTRATION: DC8-62

DIMENSIONS	Feet	Meters
WING SPAN	148.43	45.24
WING AREA square	2,927	271.93
WING SWEEP degrees	30.60	
LENGTH	150.85	45.98
HEIGHT	43.41	13.23
CREW & PASSENGERS	3/7	152

POWERPLANT		
Quantity & Type		Thrust lb/kg
4 PW	JT3D-3B	18,000	8,165

PERFORMANCE		
MAX T.O. WEIGHT lb/kg	335,000	151,954
Cruise SPEED mph/kph	587	945
MAX Mach SPEED	0.88	
Cruise ALTITUDE ft/m	31,000	9,449
MAX RANGE mi/km	3,959	6,371

McDONNELL DOUGLAS/CAMMACORP DC8-73/73F

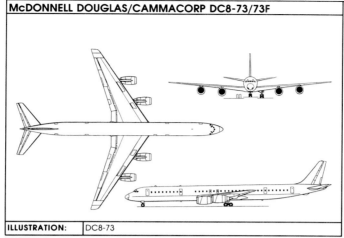

ILLUSTRATION: DC8-73

DIMENSIONS	Feet	Meters
WING SPAN	148.43	45.24
WING AREA square	2,927	271.93
WING SWEEP degrees	30.60	
LENGTH	187.34	57.10
HEIGHT	42.98	13.10
CREW & PASSENGERS	3	220

POWERPLANT		
Quantity & Type	Thrust lb/kg	
4 CFMI CFM56-2C5	22,000	9,979

PERFORMANCE		
MAX T.O. WEIGHT lb/kg	354,944	161,000
Cruise SPEED mph/kph	551	887
MAX Mach SPEED	0.88	
Cruise ALTITUDE ft/m	39,000	11,887
MAX RANGE mi/km	4,822	7,760

McDONNELL DOUGLAS DC9-10

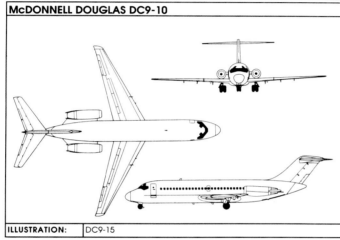

ILLUSTRATION: DC9-15

DIMENSIONS	Feet	Meters
WING SPAN	89.40	27.25
WING AREA square	934	86.80
WING SWEEP degrees	24.50	
LENGTH	104.40	31.82
HEIGHT	27.49	8.38
CREW & PASSENGERS	2	75

POWERPLANT		
Quantity & Type	Thrust lb/kg	
2 PW JT8D-5	12,000	5,443

PERFORMANCE		
MAX T.O. WEIGHT lb/kg	77,700	35,244
Cruise SPEED mph/kph	574	924
MAX Mach SPEED	0.84	
Cruise ALTITUDE ft/m	27,000	8,230
MAX RANGE mi/km	794	1,278

McDONNELL DOUGLAS DC9-30

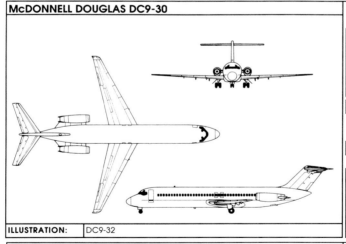

ILLUSTRATION: DC9-32

DIMENSIONS	Feet	Meters
WING SPAN	93.41	28.47
WING AREA square	1,001	92.97
WING SWEEP degrees	24.50	
LENGTH	119.39	36.39
HEIGHT	27.49	8.38
CREW & PASSENGERS	2	95

POWERPLANT		
Quantity & Type	Thrust lb/kg	
2 PW JT8D-1/7	14,000	6,350

PERFORMANCE		
MAX T.O. WEIGHT lb/kg	98,000	44,452
Cruise SPEED mph/kph	574	924
MAX Mach SPEED	0.84	
Cruise ALTITUDE ft/m	27,000	8,230
MAX RANGE mi/km	1,323	2,130

McDONNELL DOUGLAS DC9-50

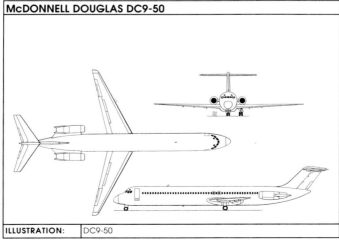

ILLUSTRATION: DC9-50

DIMENSIONS	Feet	Meters
WING SPAN	93.41	28.47
WING AREA square	1,001	92.97
WING SWEEP degrees	24.50	
LENGTH	133.56	40.71
HEIGHT	28.08	8.56
CREW & PASSENGERS	2	114

POWERPLANT		
Quantity & Type	Thrust lb/kg	
2 PW JT8D-17/17A	16,000	7,257

PERFORMANCE		
MAX T.O. WEIGHT lb/kg	121,000	54,885
Cruise SPEED mph/kph	564	907
MAX Mach SPEED	0.84	
Cruise ALTITUDE ft/m	27,000	8,230
MAX RANGE mi/km	2,204	3,547

McDONNELL DOUGLAS MD81/82/83/88

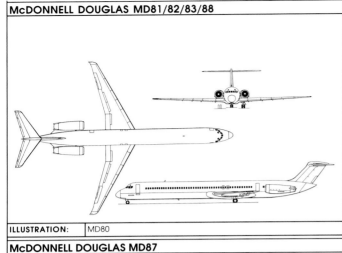

ILLUSTRATION: MD80

DIMENSIONS	Feet	Meters
WING SPAN	107.84	32.87
WING AREA square	1,270	118.00
WING SWEEP degrees	24.50	
LENGTH	147.83	45.06
HEIGHT	29.66	9.04
CREW & PASSENGERS	2	129
POWERPLANT		
Quantity & Type	Thrust lb/kg	
2 PW JT8D-209	19,250	8,732
PERFORMANCE		
MAX T.O. WEIGHT lb/kg	140,000	63,503
Cruise SPEED mph/kph	574	924
MAX Mach SPEED	0.84	
Cruise ALTITUDE ft/m	27,000	8,230
MAX RANGE mi/km	1,594	2,565

McDONNELL DOUGLAS MD87

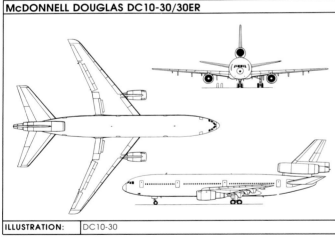

ILLUSTRATION: MD87

DIMENSIONS	Feet	Meters
WING SPAN	107.84	32.87
WING AREA square	1,270	118.00
WING SWEEP degrees	24.50	
LENGTH	130.41	39.75
HEIGHT	30.51	9.30
CREW & PASSENGERS	2	109
POWERPLANT		
Quantity & Type	Thrust lb/kg	
2 PW JT8D-217B/C	20,859	9,461
PERFORMANCE		
MAX T.O. WEIGHT lb/kg	140,000	63,503
Cruise SPEED mph/kph	505	813
MAX Mach SPEED	0.84	
Cruise ALTITUDE ft/m	35,000	10,668
MAX RANGE mi/km	2,206	3,550

McDONNELL DOUGLAS DC10-30/30ER

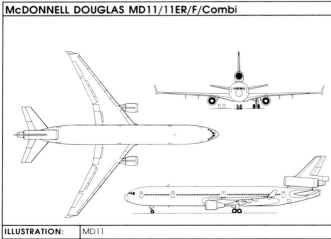

ILLUSTRATION: DC10-30

DIMENSIONS	Feet	Meters
WING SPAN	165.42	50.42
WING AREA square	3,958	367.70
WING SWEEP degrees	35.00	
LENGTH	182.09	55.50
HEIGHT	58.07	17.70
CREW & PASSENGERS	3	250
POWERPLANT		
Quantity & Type	Thrust lb/kg	
3 GE CF6-50A	49,000	22,226
PERFORMANCE		
MAX T.O. WEIGHT lb/kg	555,000	251,744
Cruise SPEED mph/kph	610	982
MAX Mach SPEED	0.95	
Cruise ALTITUDE ft/m	25,000	7,620
MAX RANGE mi/km	6,157	9,908

McDONNELL DOUGLAS MD11/11ER/F/Combi

ILLUSTRATION: MD11

DIMENSIONS	Feet	Meters
WING SPAN	169.49	51.66
WING AREA square	3,648	338.90
WING SWEEP degrees	35.00	
LENGTH	200.82	61.21
HEIGHT	57.74	17.60
CREW & PASSENGERS	2	323
POWERPLANT		
Quantity & Type	Thrust lb/kg	
3 GE CF6-80C2D1F	61,500	27,896
PERFORMANCE		
MAX T.O. WEIGHT lb/kg	602,500	273,290
Cruise SPEED mph/kph	590	949
MAX Mach SPEED	0.87	
Cruise ALTITUDE ft/m	31,000	9,449
MAX RANGE mi/km	5,409	8,704

SUD-EST SE210 CARAVELLE 6N/R

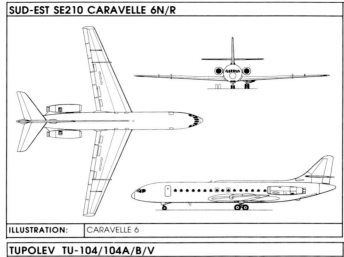

ILLUSTRATION: CARAVELLE 6

DIMENSIONS	Feet	Meters
WING SPAN	112.50	34.29
WING AREA square	1,579	146.70
WING SWEEP degrees	20.00	
LENGTH	104.99	32.00
HEIGHT	28.61	8.72
CREW & PASSENGERS	3	80
POWERPLANT		
Quantity & Type	Thrust lb/kg	
2 RR AVON Mk 531 tj	12,200	5,534
PERFORMANCE		
MAX T.O. WEIGHT lb/kg	105,822	48,000
Cruise SPEED mph/kph	525	845
MAX Mach SPEED	0.81	
Cruise ALTITUDE ft/m	25,000	7,620
MAX RANGE mi/km	1,428	2,298

TUPOLEV TU-104/104A/B/V

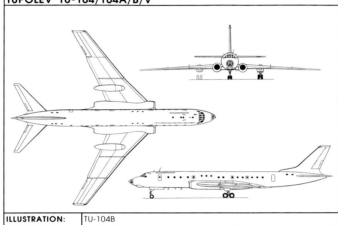

ILLUSTRATION: TU-104B

DIMENSIONS	Feet	Meters
WING SPAN	113.32	34.54
WING AREA square	1,877	174.40
WING SWEEP degrees	37.30	
LENGTH	127.46	38.85
HEIGHT	39.04	11.90
CREW & PASSENGERS	5	70
POWERPLANT		
Quantity & Type	Thrust lb/kg	
2 MIKULIN AM-3M tj	21,385	9,700
PERFORMANCE		
MAX T.O. WEIGHT lb/kg	156,528	71,000
Cruise SPEED mph/kph	590	950
MAX Mach SPEED	0.88	
Cruise ALTITUDE ft/m	33,000	10,058
MAX RANGE mi/km	1,893	3,047

TUPOLEV TU-124/124V

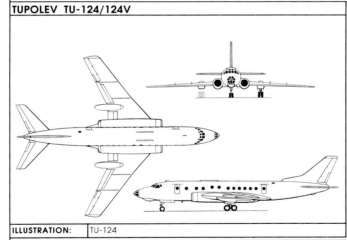

ILLUSTRATION: TU-124

DIMENSIONS	Feet	Meters
WING SPAN	83.83	25.55
WING AREA square	1,281	119.00
WING SWEEP degrees	37.30	
LENGTH	100.33	30.58
HEIGHT	26.51	8.08
CREW & PASSENGERS	5	44
POWERPLANT		
Quantity & Type	Thrust lb/kg	
2 SOLOVIEV D-20P	11,905	5,400
PERFORMANCE		
MAX T.O. WEIGHT lb/kg	83,776	38,000
Cruise SPEED mph/kph	604	972
MAX Mach SPEED	0.88	
Cruise ALTITUDE ft/m	33,000	10,058
MAX RANGE mi/km	875	1,408

TUPOLEV TU-134/134A

ILLUSTRATION: TU-134A

DIMENSIONS	Feet	Meters
WING SPAN	95.14	29.00
WING AREA square	1,370	127.30
WING SWEEP degrees	35.00	
LENGTH	112.70	34.35
HEIGHT	29.59	9.02
CREW & PASSENGERS	4	68
POWERPLANT		
Quantity & Type	Thrust lb/kg	
2 SOLOVIEV D-30	14,490	6,573
PERFORMANCE		
MAX T.O. WEIGHT lb/kg	98,106	44,500
Cruise SPEED mph/kph	559	900
MAX Mach SPEED	0.90	
Cruise ALTITUDE ft/m	28,000	8,534
MAX RANGE mi/km	1,174	1,889

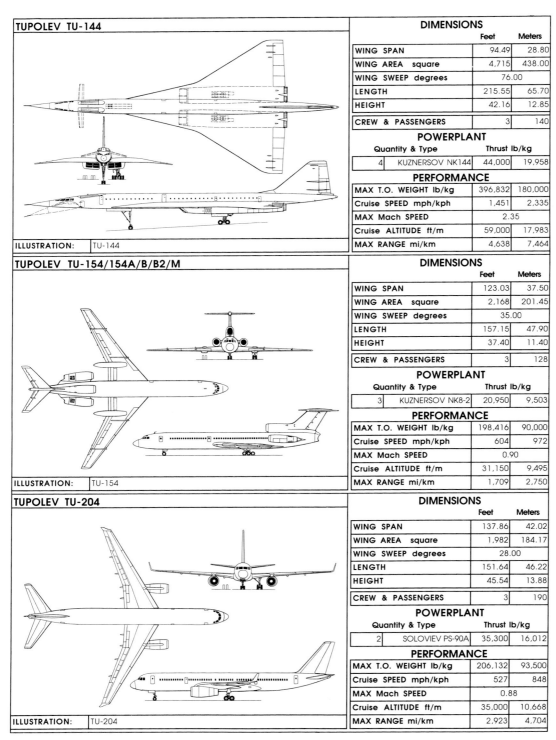

TUPOLEV TU-144

DIMENSIONS

	Feet	Meters
WING SPAN	94.49	28.80
WING AREA square	4,715	438.00
WING SWEEP degrees	76.00	
LENGTH	215.55	65.70
HEIGHT	42.16	12.85
CREW & PASSENGERS	3	140

POWERPLANT

Quantity & Type		Thrust lb/kg	
4	KUZNERSOV NK144	44,000	19,958

PERFORMANCE

MAX T.O. WEIGHT lb/kg	396,832	180,000
Cruise SPEED mph/kph	1,451	2,335
MAX Mach SPEED	2.35	
Cruise ALTITUDE ft/m	59,000	17,983
MAX RANGE mi/km	4,638	7,464

ILLUSTRATION: TU-144

TUPOLEV TU-154/154A/B/B2/M

DIMENSIONS

	Feet	Meters
WING SPAN	123.03	37.50
WING AREA square	2,168	201.45
WING SWEEP degrees	35.00	
LENGTH	157.15	47.90
HEIGHT	37.40	11.40
CREW & PASSENGERS	3	128

POWERPLANT

Quantity & Type		Thrust lb/kg	
3	KUZNERSOV NK8-2	20,950	9,503

PERFORMANCE

MAX T.O. WEIGHT lb/kg	198,416	90,000
Cruise SPEED mph/kph	604	972
MAX Mach SPEED	0.90	
Cruise ALTITUDE ft/m	31,150	9,495
MAX RANGE mi/km	1,709	2,750

ILLUSTRATION: TU-154

TUPOLEV TU-204

DIMENSIONS

	Feet	Meters
WING SPAN	137.86	42.02
WING AREA square	1,982	184.17
WING SWEEP degrees	28.00	
LENGTH	151.64	46.22
HEIGHT	45.54	13.88
CREW & PASSENGERS	3	190

POWERPLANT

Quantity & Type		Thrust lb/kg	
2	SOLOVIEV PS-90A	35,300	16,012

PERFORMANCE

MAX T.O. WEIGHT lb/kg	206,132	93,500
Cruise SPEED mph/kph	527	848
MAX Mach SPEED	0.88	
Cruise ALTITUDE ft/m	35,000	10,668
MAX RANGE mi/km	2,923	4,704

ILLUSTRATION: TU-204

Remarks on Data

The data tables provide vital statistics of all the commercial jets produced and ordered by airlines to date. The data contained is as accurate and as up-to-date as possible. All feet, miles, and pounds have been calculated from metric data.
DIMENSIONS: no inches, all are decimalized feet.
CREW: cockpit crew.
PASSENGERS: typical seating capacity for two- or three-class cabin layout.
POWERPLANT: quality, manufacturer, exact type of engine and maximum static thrust of each at sea level. The abbreviations for the engine manufacturers are:

AVCO for Avco Lycoming Textron
CFMI for CFM International
DH for de Havilland
GE for General Electric
PW for Pratt & Whitney
RR for Rolls-Royce

All are turbofan engines, unless type followed by tj meaning turbojet.

MAX T.O. WEIGHT: maximum permitted take-off weight.
MAX Mach SPEED: maximum normal operating Mach number.
MAX RANGE: maximum range with maximum payload.

Bibliography

FAA Historical Fact Book, a Chronology 1926-1971, by Arnold E. Briddon, Ellmore A. Champie and Peter A. Marraine, Department of Transportation Federal Aviation Administration, Office of Information Services, Washington, D.C, 1974

Milestones of Flight, Michael J. H. Taylor and David Mondey, Jane's Publishing Company, Ltd., London,1983

Air Dates. Air Commodore L.G.S. Payne, CBE MC AFC, Frederick A. Praeger, New York, 1957

Quest for Performance, The Evolution of Modern Aircraft, Laurence K. Loftin, Jr., Scientific and Technical Information Branch, National Aeronautics and Space Administration, Washington, D.C., 1985 NASA SP-468

The Jet Age, Forty Years of Jet Aviation, Edited by Walter J. Boyne and Donald S. Lopez, Published by the National Air and Space Museum, Smithsonian Institution, Washington, D.C., 1979

The Smithsonian Book of Flight, Walter J. Boyne, Smithsonian Books, Washington, D.C., Orion Books, New York, 1987

The Sporty Game, The high-risk competitive business of making and selling commercial airliners, John Newhouse, Alfred A. Knopf, New York, 1982

Whittle, The True Story, John Golley, in association with Sir Frank Whittle, OM, KBE, CB, Technical Editor Bill Gunston, Smithsonian Institution Press, Washington, D.C., 1987

Airbus, The Whispering European, Peter Mueller, Europa Verlag, 1984

Aeronautical Developments for the 21st Century, John M. Swihart, corporate vice president, international affairs, Boeing Commercial Airplanes, Seattle, Washington, for the 50th Wright Brothers Lectureship in Aeronautics, St. Louis, Missouri, September 14, 1987

"Airport Handling Problems to Accompany Advent of 747," William H. Gregory, *Aviation Week & Space Technology,* December 12, 1966, p. 40

David A. Brown , "Introduction of DC-9s Smooth for Delta," *Aviation Week & Space Technology,* September 12, 1966, p. 42

Herbert J. Coleman, "Autoland System Tested in Heathrow Fog," *Aviation Week & Space Technology,* December 12, 1966, p. 84

Richard G. O'Lone, "Stretched Jets to Tax Honolulu Facilites," *Aviation Week & Space Technology,* November 28, 1966, p. 47

Interview, Joe Sutter, executive vice president, retired, Boeing Commercial Airplanes, Seattle, Washington, July 22,1988

Interview, Col. Frank Borman, former president and chief executive officer, Eastern Airlines, September 19, 1988

Interview, Robert Christian, former public relations officer, Eastern Airlines and Delta Air Lines, July 22, 1988

Interview, Ralph Glasson, vice president, retired, maintenance and engineering, United Airlines, August, 1988

The Concorde Story, Ten Years in Service, Christopher Orlebar, Temple Press, Twickenham, Middlesex, U.K.; Newnes Books, 1986

General Electric Commercial Engines Program Status, Volume 1, 1987

Structural Development of the DC-10, by R. E. Bates, Orient Airlines Association., Flight Safety Seminar, Oberoi Imperial Hotel, Singapore, May 10-12, 1972

DC-10 Design Development, by R. E. Bates , ATA Engineering Maintenance Conference, Century Plaza Hotel, Los Angeles, California, December 1, 1971

DC-10 The Aircraft, Its System, Performance and Certain Options, McDonnell Douglas Corp., DC-10 Program Engineering Office, February, 1982

Airline Safety in the Last Half Century: 1950–2000 by R. R. Shaw, Sir Charles Kingsford-Smith Memorial Lecture, 1984

Paper on the Boeing 7X7, Boeing Commercial Airplanes

Problems and Challenges, A Path to the Future, J. E. Steiner, vice president-Technology and New Program Development, Boeing Commercial Airplanes, Royal Aeronautical Society, London, October 10, 1974

Aircraft and Air Transport Development, The Next Ten Years, by J. E. Steiner, Commercial Airplane Group, Royal Aeronautical Society, Johannesburg, Republic of South Africa, November 3, 1969

Systems Design for Weight Optimization: by T. P. Clemmons, research engineer, Boeing Commercial Airplanes 28th Annual Conference, Society of Aeronautics Weight Engineers, Inc. San Francisco, 5-8 May, 1969

The Boeing Model 747, A Status Report, by Jon Bedinger, Senior Group Engineer, Design Development Unit, Everett Branch Boeing Commercial Airplanes

Civil Aviation Statistics of The World, 1986, International Civil Aviation Organization

Planning for Total Support of Future Aircraft, Manufacturers' View, by Alan D. Roeser and Michael Doyle, The Boeing Co., Second Flight Test Simulation and Support Conference, Los Angeles, March 25-27, 1968

Factors in the Selection of Aircraft Equipment, R. P. Norton, Boeing Commercial Airplane Group – 707/727/737 Division

Short Range Jet Airliners Notes by J. E. Steiner, Air Transport Association of America, Engineering and Maintenance Conference, Washington, D.C., October 27, 1964

The Boeing 737 by J. E. Steiner, Seattle, April 20, 1965, Society of Automotive Engineers, Seventh Annual Air Transport Session

Interview, Clyde R. Kizer, Air Transport Association of America vice president for maintenance and engineering, August 1988

Technical paper, Evolution and Revolution with the Jumbo Trijets, Franklin W. Kolk and David R. Blundell, American Airlines, Inc., *Astronautics & Aeronautics,* October, 1968.

Airbus ... the Family Grows, *Air International,* May, 1987

The Airbus Alternative, Flight International, December 27, 1986

Airbus Picks up Speed – and the Junior A310 Takes Off, *Air International,* June, 1979

Countdown, Frank Borman with Robert J. Serling, an autobiography, Silver Arrow Books, William Morrow, New York, 1988.

Product Line Review, Airbus Industrie, October 1987

Robert R. Ropelewski, "A300 Testbed Demonstrates Fly-by-Wire Capabilities," *Aviation Week & Space Technology,* September 22, 1986.

The Air Traveler's Handbook, Simon and Schuster, New York, 1978.

Seven Decades of Progress, Aero Publishers, Inc. Fallbrook, California, 1979

"Concorde, Environmental Effects Studied," *Aviation Week & Space Technology,* February 8, 1971

High-Speed Civil Transport Studies, Phase 3, Oral Report, Boeing Commercial Airplanes and McDonnell Douglas, presented to NASA Langley Research Center, Hampton, Virginia, October 12, 1988

Index

Aeroflot 14, 90
Aeronautical Radio (ARINC) 109
Aerospatiale 57, 103
Airbus Industrie 94
   A340 9
   A300 12, 17, 41, 56, 57, 58, 59
   B2 300 58
   B4 58
   B4 200 58
   A300-600 58, 59
   A300-AFF 58
   A310 60, 61, 62, 139
   A310-200 62
   A320-100 92, 93, 94, 95, 99
   A319 94
   A340 96, 97
   A330 96
Air Canada 69, 81, 94, 139
Air France 17, 51, 58, 92
Air Wisconsin 85
Alaskan Airlines 67
Aloha Airlines 73, 139
American Airlines 42, 79, 89, 104
Avco Lycoming ALF502R 85, 86
Avro Canada C-102 14
Avro International 85
Avianca 73

Beteille, Roger 58
Boeing Company 9, 19, 34
   777 9
   707 10, 19, 23, 77, 135
   727 19, 64, 65, 66, 67, 69, 73, 90
   737 19, 73, 94, 97
   720 19, 20
   SST 37
   727-100 58, 66, 78
   737-100 72
   727-200 66
   737-200 73, 74, 97
   737-300 72, 73
   737-400 73, 75, 97, 99
   737-500 73, 75, 97
   767 77, 80, 81, 82
   757 19, 76, 77, 78, 79
   747 1st Flight 10, 37, 38
   747 81, 104
   767-300 82, 83, 97
   747-100 136
   747-200 105, 117
   747-300 105, 118
   747-400 96, 97, 105, 113
   777 97
   777-200 97
   Super Jumbo Jet (ULCT) 97
Borman, Col. Frank 58, 78
British Aerospace 85, 103
   BAe 146 85, 86, 87
   RJ 70 85
   RJ 85 85
   RJ 100 85
   RJ 115 85
   HOTOL 103
   BAC 111 105
British Airways 51, 52, 79, 83
British European Airways 30

Collins 89
Concorde 9, 51, 52, 53, 54, 105, 129, 131
Construcciones Aeronauticas SA 57
Continental Airlines 94, 133
Convair 880 19, 21
   990 19, 21, 22

de Havilland 18
   Ghost Turbojet 11, 13
   Comet I 9, 11, 13, 14, 17, 19, 57
   Comet II 14, 15
   Comet 4C 16
Delta Air Lines 21, 69, 79, 99, 123
Deutsche Aerospace 57, 89

Eastern Airlines 58, 78, 79
El Al 82
Enders, John H. 138

Fokker
   F28 88, 89
   F100 89, 135
Fujita, Dr. Theodore 137

General Electric
   CFC-50 42
   CFC-80 9
   CJ-805-3 21
   CFM-56 73, 75, 94, 97
   GE-36 99
Guillaume 11

Hawker Siddeley 13
   Trident 28
Heinkel 11
High Speed Civil Transport Study (HSCT) 103
Honeywell 89, 138
Hughes, Howard 21

Ilyushin IL-62M 90

JFK Airport   106, 107, 109, 112, 123
Johnston, A. M. (Tex)   19
Junkers Engine Co.   11

Kuznetsov NK-144   53

La Pere   11
Lockheed Aircraft Co   37
   C5   10
   L-1011   47, 49, 131
   SR-71   103
London Heathrow   111
Lufthansa   62, 73

McDonnell Douglas
   DC-10   8, 41, 42, 43, 44, 45, 101, 131, 136
   DC-8   22, 23, 24
   DC-9   69, 70, 73, 99
   DC-9-10   69
   DC-9-30   69
   DC-9-32   133
   DC-9-50   70, 74, 99
   DC-9-80   70
   MD-80   69, 71, 99, 100
   MD-87   99, 100
   MD-88   99
   MD-90   99
   MD-82   99
   MD-11   101
   MD-12   101
Malaysia-Singapore Airlines   73
Messerschmitt ME262   11
Midway Airlines   69

Next Generation Supersonic Transport (SST)   97
Northwest Airlines   79, 94, 114

Pan American World Airways   19, 63, 94, 123, 133, 137
People Express Airlines   73
Piedmont   138
Pratt & Whitney
   JJ3   19
   JT8D   65, 69, 73, 81
   JJ9D   42, 136
   PW2037   78
   JT8D-217   99
   PW-Allison Model 578-DX   99

Rolls-Royce   26
   Conway Turbojet   11, 19
   Derwent 5   14
   Avon 502   17
   RB211   39, 78, 136
   Olympus 593   52, 54,
   Tay   89
Rose, Nancy L.   138
Royal Aircraft Establishment   28

Scandinavian Airlines System   17
Short Brothers   89
Snecma
   Olympus 593   52, 54
   CFM56   73, 94, 97
Southwest Airlines   85
Steiner, John E.   74, 77
Sud-Est Aviation   8
   Caravelle   15, 57
Sutter, Joseph F.   65
Swissair   21, 22, 89, 105, 114

Taiwan Aerospace   85
Trans World Airlines   21, 82
Trident   90
Trippe, Juan   34, 37
Tupolev
   Tu-104   14
   Tu-144   53, 103
   Tu-154   90, 91
   Tu-134A   91

United Airlines   17, 74, 81, 97, 138
US Air   85, 97

Valeika, Ray   133
Very Large Commercial Transport (VLCT)   104
Vickers VC-10   26-27, 90
Von Ohain, Hans Joachim   11

Wardair Canada   61
Ward, Max   61
Whittle, Sir Frank   11
Wright Bros.   105